Science and Politics

Science and Politics

IAN BOYD

polity

Copyright © Sir Ian Boyd 2025

The right of Sir Ian Boyd to be identified as Author of this Work has been
asserted in accordance with the UK Copyright, Designs and Patents Act
1988.

First published in 2025 by Polity Press

Extract from SCIENCE AND GOVERNMENT by C.P. Snow, Cambridge,
Mass.: Harvard University Press, Copyright © 1960, 1961 by the President
and Fellows of Harvard College. Used by permission. All rights reserved.

Polity Press
65 Bridge Street
Cambridge CB2 1UR, UK

Polity Press
111 River Street
Hoboken, NJ 07030, USA

ISBN-13: 978-1-5095-6158-2

A catalogue record for this book is available from the British Library.

Library of Congress Control Number: 2024943112

Typeset in 11 on 14pt Warnock Pro
by Cheshire Typesetting Ltd, Cuddington, Cheshire
Printed and bound in Great Britain by CPI Group (UK) Ltd, Croydon

The publisher has used its best endeavours to ensure that the URLs for
external websites referred to in this book are correct and active at the time
of going to press. However, the publisher has no responsibility for the
websites and can make no guarantee that a site will remain live or that the
content is or will remain appropriate.

Every effort has been made to trace all copyright holders, but if any have
been overlooked the publisher will be pleased to include any necessary
credits in any subsequent reprint or edition.

For further information on Polity, visit our website:
politybooks.com

Contents

Preface

Not far from my home in Scotland there are enduring reminders of why we have science. These take the form of standing stones, set up over 4,000 years ago. Their precise function is still debated, but they seem to be mainly ritual or sidereal markers where people gathered to feast, socialize and recognize the rhythms of life. Archaeologists can show from the material cultures of the people who raised these stones how they used products from Nature to support daily existence as well as to build these monumental structures.

These people obviously researched their surroundings and learned how to use them, but there was much they did not understand about what they experienced, like the mineral crystals they saw in the rocks they hewed, the vagaries of the weather, the stars in the sky, the definition of life, the coming and going of the seasons, the infliction of disease and the constant rolling forward of time. This void in their understanding was filled by imagination and belief, a mixture of empirically founded theory and hocus-pocus. Even if their standing stones started as mainly functional objects for marking time, they accrued what, borrowing from evolutionary theory, we might call exaptive meaning, like a structure acquiring a function different to that for which it originally evolved.[1] We can now only guess as to what their theories and hocus-pocus amounted to because, without philosophy, the people of the stones struggled to turn their theories and hocus-pocus into structured, cumulative knowledge stored in ways which made it timeless.

About 2,500 years ago the first known philosophers started to organize and make sense of the hocus-pocus, but by then groups of people who had vested interests in sustaining and exaggerating this view of the world were in control of politics. These princes,[2] shamans and priests – whom I will collectively call charlatans – held power over beliefs and it was not in their interests to have those beliefs challenged. Ritual was invented as a way to reinforce the hocus-pocus and people were incentivized to believe that their beliefs were all that mattered. Whether this was achieved by tradition or by force, it was an assault on their liberty.

This assault continued for about another 2,000 years until the inventions of science finally allowed other forms of knowledge to escape the controls of the religious and military thought police. New knowledge became sealed in a form which could be both distributed widely and understood by many. This emancipation was powered by a technical and philosophical revolution which continues to co-evolve to the present. It defined the conditions needed to distinguish truth from hocus-pocus, and on its heels came an exponential rise in the enrichment of our knowledge about the world we inhabit, about us as animals within this world, and about our world within a wider universe – about Nature. It began to displace hocus-pocus as the font of all knowledge. This displacement continues, but the charlatans are still among us spouting their nonsense, subverting knowledge and, as a result, assaulting liberty, right here, right now.

In the present, the charlatans exist by other names, but they still dominate politics within a vastly more complex society through the application of deceptive, one-size-fits-all ideology and ersatz rituals within a variety of temples, some actual, some virtual, all designed to strip people of their free will. Their vested interests lead them to block or subvert our search for the truths around us, and this continues to cause us to make decisions which are misaligned with Nature as we know it through both sensed experience and science.[3] We are sustained by the resources of Nature, as were the people of the Neolithic, and the longer this misalignment goes on, the more painful eventual realignment will become. Continued misalignment results in the progressive magnification of unfairness and it inflicts injustice on more and more people. Scientific knowledge stops this misalignment and, as a result, it is a pathway to justice, a pathway to being treated in a fair and equitable way. Science is not itself representative of justice, but understanding how the natural world functions and how people fit in to it, which is something science has the capacity to address, is a prerequisite for a just society.

The story told in this book reflects a personal journey which began in September 2012 when I became one of the UK government's Chief Scientific Advisers. It was at that point that I began studying the ways in which science is used within government, where it comes into intimate contact with politics, ritual and hocus-pocus. It gave me an opportunity to experiment directly with different methods of delivering complicated scientific information to people at the heart of political decision-making, some of whom might be characterized as today's charlatans. Most of them were honest people, doing extremely demanding jobs to the best of their abilities. In some cases they were highly intelligent and capable, but most were struggling with the

burden of their office. Many had probably entered professional politics to try to make a difference and to do some
good, for broadly the same reasons as I had become a
scientist. But even if we are all to an extent products of
a system which rules us, in their case this assaulted their
liberty, often resulting in narrow thinking, dubious ethics
and prejudice. Some were ruled entirely by myrmidons
from the 'party' who held power without accountability.
Very few ever managed to break out of these constraints,
which sometimes drove pathological behaviours like bullying. Most cabinet ministers would have struggled to get
on the shortlist to be selected to run a large business and
yet they were often thrown into equivalent situations in
government. The worst of them were those who failed to
recognize their own limitations. Rather than being critical,
in general, I felt sorry for them, hemmed in as they were by
their politics and the hopelessness of their circumstances.

This experience was not confined to the British system;
the job brought me into contact with other national and
international science advisory and political systems, all of
which had their own charlatans. I also had been through a
career on an international stage where science was being
introduced into decision-making and, in the particular case
of the Antarctic Treaty, was the only accepted rationale for
any political involvement in that distant part of the world.

The experience at the centre of government lasted seven
years, and there were many interesting and difficult issues
throughout that time which showed me how important science is to the operation of government. Some of these are
mentioned for illustration in this book, but when I left government and started to see many of the issues at a greater
distance, without the influence of politics, personalities,
emotions and hocus-pocus, I began to develop a deeper
appreciation of what was going on. This was added to when
I was asked to provide procedural oversight for the science

advisory group set up by the UK government during the COVID-19 pandemic and to co-chair an Environment Council with the First Minister of Scotland. I also spent some time as Chair of the UK Research Integrity Office. These experiences left me with a view that the insights which science could provide were certainly under-appreciated, generally under-used and often systematically subverted, leading to bad policy and, by implication, injustice within society. I had an overwhelming sense that the pursuit of self-interest was not balanced appropriately with collective interest – in the style of Locke's and Rousseau's social contract.[4] I also saw some scientists insisting sanctimoniously that they were separate from the political process when they were obviously deeply involved. There is a lot of confusion among politicians, civil servants and scientists themselves about the utility of science as an endeavour working in the public interest and as a force for good. I also concluded that when I had started on the journey in 2012, I had been quite naïve about the political system I was entering; I have been on a steep learning curve ever since. But where did this leave others who did not have the benefit of my experience?

It is for these reasons that I have written this book. It is the book I would have liked to have read in 2012. Its objective is to improve understanding of the junction between science and politics, especially for people who have various roles within either of these domains, to help them to understand how they mingle and interact and how this can perhaps be made to work better. I believe those who are practitioners in the domains of science and politics have a duty to improve and, likewise, I have a duty to share my insights with them. I also hope this book may contribute a better understanding among many people who simply observe this interaction from a distance. When done well, the delivery of science for public good can improve the wealth of nations in ways that far surpass simply adding percentage points to soulless economic indicators.

I also had the aim of helping to debunk some common myths and misconceptions – such as science being an apolitical endeavour, or scientists somehow being more honest than politicians. The biggest myth of all is that scientists themselves cannot take part in politics, something I once thought was true, but perpetuation of this myth simply leads to worse politics, grounded in pure emotion and ideology, and to a lot of scientific research which is either completely subverted by politics or is irrelevant to public interests.

I am a natural scientist who has spent most of his career exploring some esoteric topics, mainly in animal physiological ecology and polar and marine science. But this means I can bring a unique and, hopefully, balanced, blended perspective about a topic which can raise some people's hackles and result in a lot of blame shifting. The messages from my own knowledge of how systems in Nature operate, and sometimes fail, are a constant reminder that as organisms governed by the exact same physical forces, humans are not immune to these operational constraints. What has become clear to me is that, if people confine their appreciation of how society works to society itself, to what I call the *social bubble*, and forget about how society is ultimately constrained by Nature, then eventually only injustice can prevail.

I use 'hocus-pocus' when referring to systems of decision-making and rules for living which are based in blind, absolutist belief. It is 'blind' because much of it has been built without reference to the realities of Nature. It is often deceptive, and even when it might be correct, it can be deceptive about why it may be correct. Hocus-pocus is also dangerous because those who promote it fail to ask questions about its validity and to modify it based on experience.

There is at least a little of the charlatan in all of us, but when we ask these questions, we become scientists. Even if this book makes us stop for a moment to think twice about what we are doing, to readjust and to modify our behaviour, then it will have succeeded.

The book tests four ideas. These are, first, that science is inherently political; second, that the structures within politics, which date back to before science was recognized as such, actively exclude scientists from the political debate; third, that this exclusion leads to injustice; and, fourth, that things can be done to resolve this problem, but they require change both within science and within political systems. My contention is that justice is unlikely, perhaps impossible, while science is not a fully integrated part of the systems for collective decision-making across society.

The book is structured in three sections. The first section considers the state of the current relationships between science and politics. It explores the old but still popular notion of science and politics representing a clash of cultures. It characterizes the relationship as a marriage where there is co-dependency in a partnership but also examines politics as a societal process of debate and collective decision-making which is a trade-off between public and private interests[5] and of the dilemmas this creates for scientists. I use the metaphor of a 'factory' to describe the process by which politics happens because politics can be characterized as an emergent product constructed from many components. There may be no single guiding mind or design, but its products take the form of laws, regulations, agreements, contracts, conventions, protocols and policies, as well as debates, arguments, wars and conflicts.

The second section considers the impact of politics on science itself. If science is the main way in which we introduce objectivity into the operation of society, then it is subject to many different corrupting forces. Science struggles to move beyond simplifying platitudes like 'evidence', 'what works' and a hunger for 'experts', especially among the news media. I call this process the subjective by-pass. It is a structural problem which leaves our decision-making in the domain of public policy in what Rachel Carson called a 'Neanderthal age of biology'.[6] Distilling quality in science out of all the corrupting influences is a rare skill which turns scientists embedded within political structures (like I was) into graven sceptics.

The third section is more uplifting and optimistic. I use the metaphor of a 'beast' to describe the kind of organic structure which science is attempting to interact with and to influence. The body politic is like a large, complex organism which has behaviour emergent from formally complex underlying causation. This involves lots of features of systems where there can be many stabilizing feedbacks, most of which make it very difficult for anything to change as a result of single, small-scale interventions typified by the ambitions of politicians and their policies. One could fume or despair at the belligerence of this beast but, looked at with the eye of the biologist, the question is how one designs processes to shift and manage this organism in a manner which leads to greater justice.

In the final chapter, I suggest the kind of bureaucratic engineering which I think is required to mix technocracy with democracy and to avoid plutocracy.

The book is, therefore, emphatically not about creating and delivering a manifesto for geeks. Others have tried this[7] and, apart from failing to make any difference, such an agenda just establishes science as a kind of ideology[8] to be lost in the din created by yet another group of people with a special interest whose main aim is to carve out a bigger slice of the cake for themselves. Some scientists would have science trying to shout louder in a noisy room and others, with all modesty and humility, explain that their voice is just one among many. With respect, I doubt if either is right. Science is too important a 'conscious artefact of mankind'[9] to be caught up in the parochiality of fighting to be heard. Science cannot be just another thing which people do, like plumbing, accountancy or teaching. To see it this way would be to deprive people of the knowledge they need to rid themselves of the shackles of hocus-pocus.

This is neither a work of philosophy nor a work of social theory. Even if I refer on occasion to both, I want to be clear that this is a work founded in empirical experience. I am an empiricist at heart because there is nothing quite like the experience of living and working at the coal-face to understand its structure. To some extent this makes it a counterpoint to much social science in the field of science–policy interaction, where there can be too much theory and too little practice. It also impinges on political science, but it is not meant to be a work in this field either. I would not consider myself qualified to say anything original in these other disciplines, but it may be that the perspective I am able to give is helpful to those who focus on them. It would be impossible in the space of a few hundred pages to recapitulate the deep and important philosophical journey which has brought civilization to a point where science, as a rising power in the world, needs to be viewed with greater clarity and understood and used with greater depth of wisdom. As I write, some people fear we have unleashed a ravenous, destructive force on the world in the form of artificial intelligence. We need to get a lot better at governing our own innovative capacities or else these will end up supplanting the old hocus-pocus and do as much or perhaps even more damage to justice.

This is also not a work which is designed to illustrate the wonders of science. I am not keen on the glossy illustrations of science given in the popular press and media, where it is touted as some sort of playground for geeks or a crystal cave

full of glittery, wondrous facts which are largely irrelevant
to a secular audience. My experiences in government – at
the sharp end of science – have only magnified my appreciation of the distance between the public perception of
what science actually does for us and what people think it
does for them. Science is full of hard, sometimes unpleasant
lessons for us all, and turning away from them, presenting
only the crystals from the cave of wonders, does us no good
whatsoever.

For some scholars this book is inevitably going to be full
of holes. If these people see it as an attempt to re-set how
science works within politics, then they go too far. It is
one perspective only. A lot of the questions I found myself
asking as a result of my experiences are not well illustrated
in social or political theory, or, even if they are, they are
largely inaccessible to non-specialists. If I have missed some
crucial aspect of theory or argument, then I can only apolo-
gize. I am trying to speak to people who are not steeped in
the subject matter of the sociology of science and technology but who, like me, just want to know what it is like in
practice and who might benefit from hearing a voice which
has been through the mill of experience.

One lesson from this experience is that I see a need to
shake many people out of their unquestioning acceptance
of existing perceptions. My conjecture is that there is a
ridiculous over-abundance of normative behaviours and
this has left rational scepticism on the sidelines leaving the
way clear for scepticism to be shaped by people for whom
science is anathema and for whom rationality stems from
pure belief. Moreover, those people who promote the cults
of 'evidence' and 'what works' create the conditions under
which science-free policy can flourish under the deceptive guise of evidence-based policy. I appreciate that many
well-meaning people might find this absurd, perhaps even
offensive, but my purpose is to suggest they stop and think

more about what they are doing because their efforts often achieve precisely the opposite of what they intend.

A word also on terminology. I am painfully aware that in order to summarize and avoid getting bogged down in complex explanations, I use terms like 'scientist', 'politician' or 'people' as very broad categories. These are, of course, diverse sets, and nothing in what I say is meant to diminish this diversity or demean those who might not fit more or less into any particular categorization. These categories are less about pointing to individuals and their values and more about behavioural categories that we all adopt in different circumstances. For example, I argue at one point in the book that we are all 'scientists' to an extent. In the same way, people who self-identify as 'scientists' are also 'citizens' and can also behave like 'politicians'. I have also tried to avoid technical terms but, when used, I provide a definition as an endnote. I have also intentionally avoided building the arguments around personalities and events. Besides having no wish to criticize specific individuals or betray confidentialities, involving personalities also subjectifies issues which are actually more often reflective of systemic challenges. While individual personalities can be influential in shaping issues, in general, they are no more than actors or nodes in wider systems.

The sense in which I refer to science is captured by the German term *Wissenschaft*, which is simply the pursuit of knowledge, learning and scholarship, a pursuit of general truth. Science is a system of knowledge but it is also a process for observing and helping to understand the structures and functions of Nature, which happens also to include humanity and the interests of people.

My use of the term 'Nature' also needs to be understood. It is not resonant of some aesthetic ideal (as one reviewer of this book seemed to think); rather it is the foundation of the realities which constrain and contain our existence.

It is all matter from fundamental particles like bosons and fermions all the way to the aggregate structures like stars and galaxies within the cosmos and everything in between, including the contents of the biosphere of planet Earth of which we are an integral part.

I have provided references within the text where there may be a need to support statements or to help readers follow up with additional research, but these are not designed to be comprehensive. I have also used italicized text to provide emphasis and to highlight key terms when they are used. Where appropriate, direct quotations have been referenced.

Finally, I am grateful to the reviewers of this book and several other readers for their helpful comments, to my wife Sheila for her support and comments and the editorial and publishing staff at Polity Press for their enthusiastic encouragement. I am also grateful to colleagues from both academia and government, who are too numerous to name individually, for having provided support and challenge over many years. In particular, senior colleagues at the University of St Andrews never seem to have doubted providing me with their unconditional support.

Introduction:
The scientific predicament

During the COVID-19 pandemic, Patrick Vallance, the Chief Scientific Adviser to the UK government, said, 'I've got one piece of advice for any science adviser, it's: stick to the science.'[1]

Some people say that science and politics should not mix,[2] but I suggest the opposite is true: not only are science and politics closely enmeshed with one another and becoming more so with time, but this is a good thing because it is likely to result in better politics, and maybe even better science. Yet how can we reconcile this with Vallance's advice?

First, we need to understand what politics actually is, which is something I suspect remarkably few people stop to really think about. In my view, politics is the manifestation of how we negotiate our ways through life as social organisms. Even as spectators of the high politics happening within national democracies, or even autocracies, we are participants. Politics is how we deliberate about making collective decisions.[3] It is the process we use to rub along. It involves those who have a voice explaining their own interests and also listening to other perspectives. This happens at all scales, from households to the floor of the United Nations

General Assembly. Rules of procedure – some formal, such as voting, and some socially modulated, such as consensus forming or even various forms of coercion – are then used to come to a collective decision. Once this collective decision is made, then it can be encoded as policies which are statements of collective intent. These provide guidance for those who are allocated the duty to implement decisions derived from the collective view. Some of these intents end up as laws or regulations, and it is this process of deliberation which mainly shapes the societies we live in.

But politics can also involve lots of mendacious behaviours to shift the balance of power within these debates, and I think this is often how it is viewed. We are all politicians, but many of us do not like to admit this in case it exposes our own mendaciousness, however trivial that may be. Of course, professional politicians have no such defence. Politics exaggerates behavioural pathologies associated with vested interests, tribalism, selfishness and the traits which make humans both successful and rapacious.

It seems to me that scientists need to be intimately involved in the process of deliberation which underlies all politics because without their input some of these pathologies can go unconstrained. Many people might say (as did Solly Zuckerman,[4] a former Chief Scientific Adviser to the UK government) that if scientists want more political influence, then they should use the democratic processes by putting themselves up for election. This seems reasonable in principle, except for two caveats. The first is that current political structures require rationalists like scientists to become hooked to a political party, which is equivalent to being hooked to an ideological position. It is, therefore, obvious that such a proposition is absurd. Current political structures are segmented by ideology and simply cannot accommodate people whose wish is to cut across all ideologies by championing objectivity.

The second caveat is that these critics are reinforcing a very narrow but all too prevalent definition of democracy by imagining that power only flows through elected representatives. Democracy is more than just about being elected to representative assemblies like parliaments. The institutions of democratic societies, from legislatures to the judiciary and the executive, plus all the paraphernalia of public, commercial and third-sector bodies surrounding these, many of which provide essential services across society, are just as much a part of the democratic process.[5] Science can easily play its role through these, and often does. For example, science happening within government, academia and industry worked synergistically to solve the COVID-19 crisis. Government science provided public health surveillance; academic science came up with new vaccines and diagnostics; and industry science took those ideas and rapidly scaled them up making them available to everybody. This was an illustration of how science can operate effectively in a democracy without the need for scientists to stand for election.

Even if it showed what can be achieved, however, the COVID-19 experience was an exception in terms of the constructive mixing of science with politics. Making this the norm will require people to ditch some old ideas, and this applies as much to scientists as to politicians. These old ideas involve the suggestion that science is not about values and seeing politics just as a process practised by a few people – sometimes generically and stereotypically referred to as 'decision-makers' – rather than everybody.

It is a common stereotype to see politics as concerning values whereas science concerns facts. This is something political scientist Leo Strauss called the *fact–value distinction*,[6] but he concluded that claims about the value-free nature of science are bogus and toxic. We need to openly acknowledge that scientists always bring values to any argument; otherwise facts and values become blurred and we lose our bearings in the resulting fog, something which the new age of disinformation aptly illustrates.

There can be no better illustration of the toxicity of the separation of facts and values than the role which science played within Nazi tyranny. Carl Schmitt, a political theorist who was a prominent member of the Nazi Party, promoted the idea that politics exists independently of science. This justified disengaging the morality of politics from the analysis of its consequences. It allowed falsehoods to bloom as if they were truths and, in the present day, it is part of the playbook of the proto-tyranny of populism and the idea that science functions simply to amplify narrow political objectives like economic growth and competitiveness.

Other prominent political scientists of the 20th century, including Max Weber, but also Hannah Arendt and Hans Morgenthau (both, like Strauss, refugees from Nazi tyranny), saw an additional problem concerned with applying science to politics by reducing politics to rational paradigms and frameworks, essentially turning politics into technocracy.[7] Weber saw politics as 'a strong and slow boring of hard boards' involving 'passion and perspective',[8] but he also emphasized that it had a practical, dispassionate, pragmatic side which prioritized reality over imagination, and science provides the bulwark for this earthy aspect of politics.

Facts – the domain of science – and values become merged within politics. So politics is something more profound than pure passion or aspiration and science is more than just paradigms and frameworks: they both represent the hard work of making society function by merging reality with aspiration. Together, they are about establishing the rules of the game of life to achieve a balance between collective and individual good. Much political debate becomes centred on where this balance should sit, and science has a legitimate role in this debate. It can be the vehicle deployed to satisfy aspiration by driving the process of discovery and invention, but it also tends to observe that not all aspirations are feasible. When these messages become portrayed as restrictions of liberties, science can be unfairly viewed as a left-leaning conspiracy. It then becomes a target for those who think that re-engineering the uncomfortable messages from science, or ignoring science altogether, is the fairest way forward.

One constraint on making Weber's merger work is that *public* politics is necessarily a highly abstracted (and in many ways deeply dishonest) depiction of what happens in *private* politics, where there can be genuine efforts to include science in deliberations about policy. The predicament faced by politicians is that they cannot afford to tell those who vote for them that many of those voters are bigots who fail to recognize that life is full of trade-offs and that compromise is essential. This means politicians can look like liars because, to please the bigots, they get trapped into making undeliverable promises. Of course, some politicians are themselves bigots, but it is the general fear of being exposed as such, and as liars, that mainly makes them wary of scientists.

Therefore, those who suggest that science should not mix with politics are making the case that science should not inform the collective decisions which lead to the codes by which society functions. The mistake they make, I believe, is in equating politics with the passionate spinning of dreams rather than Weber's 'boring of hard boards'. As a result, either they tend to join these people in the spinning of dreams or they walk away and allow the dream spinners to have a free run. Neither seems right.

A basic principle of creating tolerance and common understanding is to connect rather than to isolate differing parties, and scientists need to promote the making of connections.[9] But this requires an especially energetic approach by scientists. In an essay written in 1945 called 'The Evil of Politics and the Ethics of Evil', Morgenthau began with the statement 'Man is a political animal by nature; he is a scientist by chance or choice.'[10] He saw science as something which was not innate, meaning that scientists are always having to educate, re-educate and then re-educate again. Moreover, Morgenthau thought that this was essential for ethical existence and that without a lot of hard work by scientists, among others, politics eventually descends to tyranny.[11]

Scientists cannot, therefore, disengage with politics, because when this happens, the foundations for ethical existence disappear. However, while this explains *why* scientists need to engage with politics, it does not explain *how* to undertake such engagement.

Religions have similar dilemmas, and in this context Strauss identified what he called the *theologico-political predicament*.[12] It is often exemplified by the syndrome of politicians railing when bishops speak out on political issues. The predicament is caused by separating ethics from politics by applying the convenient excuse that we need to avoid theocracy rather than submitting to a proper debate about the morality of certain ideologies.

The homologue for science would be, I guess, a *scientifico-political predicament*: the attempt to separate science from politics to avoid technocracy. The predicament, at least for scientists, is that if they get involved in politics, they end up becoming corrupted. Saying 'stick to the science' is one way to avoid this, but it is deceptive about the values carried by scientists and the value which science can bring to politics.

Various conventions have arisen to try to address this predicament. For example, scientists often refer to themselves as *impartial* and *independent*, but these terms are loaded with caveats. They are mainly about how scientists like to portray themselves and, certainly, many politicians

I worked with saw scientists as neither impartial nor independent. When being viewed by somebody who is steeped in certain beliefs, one is neither impartial nor independent if one is a champion of objectivity. But many scientists also use these terms to hide their real ideological affiliations.

Being apparently humble is another suggestion,[13] but this is not a neutral position, especially when this is confused with advice that scientists should maintain a 'professional distance' from politics. There is an important difference between avoiding getting involved in ideologically based arguments, which is what many commentators are rightly referring to, and counteracting the temptation by being distant. Informing has never been promoted by disengagement, yet this is how such urging is often interpreted.

Scientists – or any free-thinking intellectuals – probably have four modes of behaviour to choose from when operating alongside politics. The first is complete disassociation, to remain distant and other-worldly, to maintain a 'professional distance'. A second is to engage either by prognostic opposition to almost all politics, by opening the can of worms for all to see, or by sanctimoniously speaking truth to power. The third mode is for scientists to surrender their freedoms, becoming captured by politics, even becoming apostles of politics, and cogs in the machine. This, as I shall explain, is where a great many scientists end up, even if they do so unwittingly. The fourth, and most difficult, mode to pull off is to collaborate but only up to a point. Rather than speaking truth to power, this mode integrates truth with power by operating inside the mechanisms of politics. As a result, it accepts the restrictions which come with this and the need for pragmatism and compromise. I was operating in this mode as a Chief Scientific Adviser. People operating in this mode have been variously described as 'honest brokers' or 'knowledge brokers'.[14] Sometimes scientists also operate in the role of issue advocates when they are involved in promoting specific knowledge, but none of these terms truly capture the richness of the relationship needed when science operates as a strategic guide within politics. In the end, however, none of these modes is really a satisfactory way of managing the *scientifico-political predicament*.

The predicament lies in the bigger picture presented by thinkers like Strauss and Morgenthau, who anticipated that, because of the ways it explains reality, science would become a force within politics. By showing us that power ultimately lies in Nature, science reminds us that the hyperbole of politics is ephemeral. As the holders of the keys to our understanding of Nature, scientists, therefore, hold the keys to power. This makes them both important and the potential target of all sorts of subversive activities, creating a bewildering predicament for scientists about how to wield their power. Their disengagement from politics becomes the simplistic solution.

As one scientist who reviewed this book said, 'As a scientist I seek universal truths. Politicians lie to achieve short-term goals. This benefits the politicians but not society. Hence I have never wanted to engage with politics.' This is a fair and understandable position, but it is equivalent to handing the key to power to people who probably cannot be trusted to use that power wisely. As Plato said, 'Those who take no interest in politics are doomed to live under the rule of unworthy people.'[15]

The deep dilemma this creates for scientists is illustrated by an experience I had with one of my colleagues who was writing advice about his research for some government officials and was very reluctant to move beyond the purity of the facts at hand. I advised him by saying: 'You know more about this subject than anybody else on earth. Why should you not tell them what you think they should do?' The problem faced by him, which Morgenthau also recognized, is that the very act of acting potentially makes you immoral because there is a chance you might be wrong. But if scientists always position themselves as passive observers, only intervening when others think to ask for their advice, this is potentially equally immoral, especially if it deprives people of the knowledge they need to live better lives.

When among the politicians I advised, I was often thankful I was only the adviser, but should I have shared some of the responsibility for the decisions being made? If I had shared responsibility, would I have tended then to provide advice which was protective of my interests rather than built on objective analysis? For these reasons, allowing scientists to provide advice without accountability for that advice is a central tenet of current scientific advisory systems, but is this right? Juliet Gerrard, a Chief Scientific Adviser to the New Zealand government, seems to think so. She made the point that, 'Science is never the only advice. Science can help inform the decision-makers, [but] scientists are not making the decisions.'[16] I am not so sure. Politicians may be the fall-guys who shoulder the accountability when things go wrong, thus saving those elsewhere in the decision-making system from having to protect their own interests, but decisions in government are rarely a matter of a choice made by a single individual. Instead, they integrate among many influences, so in my view it possibly amounts to abrogation to suggest that these influences, and the influencers behind them, are not also making decisions. I shall refer later to how this plays out in particular when it comes to how technical analyses known as *impact assessments* are used to direct policy decisions.

I want not just to help scientists themselves to resolve their predicaments but also to illustrate the problem to others who might be spectators. Ignoring politics is, therefore, not an ethically robust option for scientists. That being said, I want to emphasize that the solution I propose in this book does *not* involve shifting science closer to politics. We will only be able to solve the *scientifico-political predica-ment* if politics shifts to resemble science.

The COVID-19 pandemic has had all sorts of impacts. Most of these have been bad, but there might be one legacy which could bring lasting good: perhaps it has helped people to better understand that they are vulnerable.

A more subliminal effect might be a lasting appreciation that it is science which both describes these vulnerabilities and helps to protect us from them. It was vaccine technology which really solved the problem of COVID-19, aided by better therapeutics and knowledge of epidemiology. But it was also science which brought the hated 'lockdowns' and 'social distancing' as interim measures to reduce the impacts of the disease.

Science also tells us that COVID-19 could have been much worse than it was and that other, similar events are very likely to happen even within the next five years.[17] A pandemic of a respiratory infection was inevitably going to happen, and the fact that it turned out to be a virus which killed only 1 per cent of its victims was about as good as we could hope for. When COVID-19 struck, it was an unknown virus, but a close relative, SARS-CoV-1, threatened to become a pandemic in 2003, and it killed nearly 10 per cent of those infected. Another, known as Middle East Respiratory Syndrome, or MERS, killed 35 per cent.[18] Neither of these other two viruses was as transmissible as COVID-19, and that is what saved us from them, but this happened by pure chance. It was Albert Einstein who said that 'God does not play dice,'[19] but he was wrong: the dice that rolled in 2019 when chance modifications happened to make the SARS-CoV-2 virus transmissible in humans was a relatively lucky throw.

What would it have been like if MERS had been as infectious as COVID-19? We can be reasonably sure that when the level of sickness in a population leads to an absentee rate of around 25 per cent, then things we take for granted start to fail. Food stops being delivered to supermarkets, water and electricity supplies start failing too, as do hospitals and the delivery of drugs to pharmacies. People start to die, not of the disease itself, which is bad enough, but because the support structures they need for life start to disappear. Government also starts to fail because those people we all rely on to keep the wheels turning behind the scenes also get sick. What we generally describe as 'the economy' no longer exists. Society itself starts to fall apart. One exercise I was involved in which simulated a highly pathogenic infectious disease came to an end when we could no longer manage the dead, let alone the dying. And what would emerge from the mess would probably not be the current kind of democratic politics built carefully over centuries of struggle. Instead, it would be back to the Hobbesian warlord politics of 'all against all'. People are oblivious of just how close we are to this kind of dystopian future and how much goes on behind the scenes to stop this happening.

Glimmers of these kinds of problems appeared during the COVID-19 pandemic.[20] The public was only vaguely aware of them because of shortages of a few items in shops and because of the appeals from government to stay at home and socially distance during the pandemic. Libertarians, who often appeared to prioritize ideology ahead of saving lives, made the case that sustaining the systems which ran society also saved lives in the long run. Fortunately, in general, countries managed to keep essential services and activities going, but we know it was a close call. A problem for governments is that systemic collapse is not a gradual process: it is more like the collapse of a house of cards, sudden and powerfully transformative with almost certainly no option for recovery. In science, we call it a bifurcation or state transition. In more basic language, it is like falling off a cliff.

The probability of such an event is quite high in view of current trends of population pressure, resource use and stress on the environment, all of which add up to us levering more and more out of the systems which sustain us. These systems do not have infinite resilience, even if we tacitly assume they often do. They are also not always inherently stable, or 'equilibrium' systems, as many people assume. The probability of system bifurcation is, therefore, not the same as a rare event like a large asteroid hitting the Earth. There are many known viruses which could create a much worse pandemic than experienced with COVID-19 and many others are currently unknown. The risk of nuclear war is high and, even if a war ends up being just regionally based, it would be very likely to result in rapid failure on a global scale of most of the mechanisms we rely on for normal life as we know it. It should be obvious to most people that these kinds of events would only have to happen once and we would be in dire trouble. One would have thought, therefore, we would be highly sensitized and precautionary about going anywhere near these kinds of scenarios. Yet we think little about these things and almost never plan for them. The experience of COVID-19 also showed us that even if we were to plan for them, our capacity for operational delivery of those plans can be very limited. If plans rarely survive contact with the enemy, then the best tactic is to avoid contact with the enemy altogether.

These are tough messages, but some people, like me, think about this kind of stuff and try to do something about it. We are a group of scientists who are engaged with defining these challenges, searching for solutions and attempting to ensure that people in positions of power understand the risks and, wherever possible, act to reduce them. The fact that bifurcations have so far been avoided perhaps suggests some level of success.

Most of these people hardly appear in public, but during the pandemic they became public figures. People like Anthony Fauci, who as Chief Medical Adviser to the President of the United States was one of very few to defy Donald Trump and not get sacked. In the UK, Patrick Vallance, the UK's Chief Scientific Adviser, stood beside and in support of the Prime Minister in daily briefings on the pandemic, as did Chris Whitty, the UK's Chief Medical Officer. There were, of course others elsewhere, like Anders Tegnell in Sweden, who was the architect of Sweden's novel approach to pandemic management involving, at least during its early phase, placing more trust in people to instinctively shift their behaviours to reduce transmission risk while going about their normal business. The politicians in most countries had their advisers like these people, and the World Health Organization had the role of advising internationally.

Most of the time, these people operate behind the scenes, some inside government but many outside. If these people have power, it comes from their capacity to explain the reality of how the world works outside the purely social relationships between people. They help people to understand when beliefs are aligned with reality and when they are just hocus-pocus. One of the great lessons from occupying these positions is that it is one thing to give advice but it is quite another to ensure it has been understood. Even intelligent people have ways of avoiding being confronted by reality. The Latin proverb *magna est veritas et prævalebit* (great is the truth and it will prevail) has a strong presence in their work, even if often they realize that making truth prevail can be tortuous and great harm can be inflicted in the time taken to get there. They are not politicians hunting for votes or people on a quest for self-aggrandizement, which means that their messages (whether delivered behind closed doors in private to politicians or to the public) can carry significant weight.

For example, my favoured message when anybody asked was that we face a lot of challenges over the next few decades at a scale and with possible impacts which have never been experienced before in recent human evolution. Some of these challenges include non-communicable diseases (such as obesity); others relate to exotic infectious diseases, including viral diseases but also increasingly bacterial diseases (because of the challenge we have as a result of anti-microbial resistance to medicines and cleaning agents); but others still, which are even more concerning, involve the increasing rate at which we are using global resources and the consequential rise in global pollution as these resources are processed and spat out by our economy. Even food production, which we mostly treat as a renewable resource, will decline because in its current form it is generally a process of mining soils for their minerals. Some of the critical minerals, such as phosphorus, needed for plant growth, will eventually run out.

When one considers the undoubted relationship between the rise and fall of greenhouse gases in the atmosphere and the waxing and waning of glacial and inter-glacial periods over the past 2 million years, one cannot but be concerned about the extent to which humans have changed the atmosphere of the planet to such an extent that we now live in times of unprecedented atmospheric composition. This means the benign climate which has existed for much of the period over which humanity has flourished could shift very rapidly indeed, and whatever way it shifts will be difficult for a very large proportion of the human population. We have, in addition, reached the productive capacity of the planet for food using the technologies of today (such as ploughing fields and scattering seeds or reaping the ocean for its fish and krill).[21] We are starting to feel the pinch as a result of some of these resources becoming scarcer, but it is the waste we produce from this resource consumption which is starting to affect us even more than resource availability. This will create a vicious cycle – a feedback loop – causing more non-communicable diseases, and more distress. As an example, we experience this because of the effects of vehicle emissions on air quality, something which both debilitates and kills people. But endocrine-disrupting chemicals, which we emit into the environment, are also a problem which is just one small step away from seriously affecting people.[22]

All the resources we consume and which enter our economy or our bodies, in the case of food, have to be excreted in some form as waste. Greenhouse gases and the story of global warming are a part of this picture, so the focus on eliminating fossil fuels – the 'net zero' objective which has become so prevalent – is only one part of a need for a broader and much more demanding objective involving the need to reduce resource consumption. People like me consume on average about 18 tonnes of raw material per year, and this needs to get to below 8 tonnes if we are even going to have a hope of sustainable existence.[23] Some of these challenges will be tackled through a combination of technological solutions and conscious and deliberate management. Relying solely on the headlong hedonism of the market will only lead to a struggle to contain rampant consumption. In the language of the economists, we need to balance supply-side with demand-side solutions. Rather than incessantly ramping up supply, we need to constrain demand.

These kinds of messages can be tough to take, but they are the realpolitik of our times. They should make us pause to think every time we take the trash out, or decide to get in a car, or just to buy anything, whether physically manufactured goods or a service of some sort. Instead, we are programmed to put social need ahead of dealing with these kinds of existential problems. We are like lemmings running headlong towards the proverbial cliff edge, and it can seem like there is nothing that can be done to stop the inevitable tragedy. Despite all we have learned about the rise of exotic diseases from the COVID-19 experience, the updated US National Biodefense Strategy in 2022 glossed over the idea that new diseases could emerge.[24] We can quickly forget that the policies we create and the laws we pass to oil the wheels of society depend for their robustness on the Laws of Nature, which has no respect for our laws and which will find and exploit every weakness in them.

Part of my job when I was a Chief Scientific Adviser in government was to reinforce the understanding of this dependency on Nature. It was to remind the political class that true political power lies with Nature, and that if we forget this, then Nature will impose its will on us in ways which will be brutal. COVID-19 was only a small taste of that brutality, and if we learn from it, then we are better off. Science, then, has much to say about most (perhaps all) political decisions, although it is often treated as something separate and distant from social reality. Many politicians see it as a mechanism for fuelling the consumption machine rather than slowing it down and many scientists happily acquiesce to this view.

COVID-19 was an avoidable tragedy. Many scientists, as well as early warning systems which existed within many nations, were signalling that something of that ilk was going to happen. What went wrong, then? The superficial reason for this is that government systems of surveillance for the kind of risks which could have disastrous consequences were flawed. They sent the message to politicians that the risk of something like a pandemic was much lower than it really was.[25] A lack of instinctive understanding of probability by those in power and their advisers and civil servants contributed to this. Ministers and civil servants struggle when confronted by the scale of the challenge of the risks they are responsible for managing. By failing to communicate them to people, they fail to share these risks in ways which would result in societal change and adaptation to reduce them. As a result, people live almost entirely in a social bubble insulated from having to think about how they might be affected if a pandemic (or some other existential challenge) should materialize.

To illustrate, in normal life when we perceive risk we either do something to mitigate it or we can take out insurance by sharing our risk with others in a similar position. However, governments are normally responsible for insuring themselves but they fail to do the job of insurance companies, which assess the probability of bad events and then price these into insurance premiums. As the insurers of last resort, governments stepped in when banks ran out of cash in 2008, when fuel costs rocketed in 2022 and when furlough schemes were created in 2020 to stop mass unemployment owing to the collapse of market economies throughout the world. Unlike insurance companies, governments never add up their overall risk and build this in to their financial and social planning. In general, they swan on, hoping that bad things won't happen, and when bad things do eventually happen, they hope there will be enough resilience in their systems to adapt and cope. Some investments, like those in defence and healthcare, could be characterized as a form of insurance but, in general, there is no explicit understanding of the liabilities carried by governments or of national resilience.

This means that planning, policy development and budgeting generally fail to account for oncoming disasters, and this is really why the COVID-19 pandemic happened. A focus on economic competitiveness, rather than cooperation, leads to a toxic situation in which the trade-off between investing in long-term resilience and investing in short-term economic growth becomes skewed more and more towards short-term needs, which are what attracts votes. To use a simple example, in the months before the pandemic struck, a politician would have attracted little public support for choosing to invest in respirators and personal protective equipment (PPE) to be stockpiled for some future medical emergency when the money used could have been invested in kidney dialysis needed immediately. This kind of process drives the vehicle of government faster towards the precipice of bifurcation rather than slowing it down.

For all these reasons, one of my last acts when in government was to refuse to sign off on the UK's national risks assessment because I thought it was deeply flawed. I recall the anguish this caused among civil servants, but this was six months before COVID-19 struck and I knew we were cruising towards some sort of disaster. I just did not know what it would be or when it would strike.

In the years before the pandemic struck, people like me were pointing out this dangerous miscalculation. But top political priorities were, and still are, hedonistic and narrowly conceived economic growth.[26] Few politicians on any side of the ideological spectrum are serious about their duty to manage risk (except perhaps electoral risk), but achieving better integration of scientific insight with politics could help to re-balance priorities. The strategic role of science in helping to re-balance priorities was what Robert Oppenheimer was calling for in his 1953 Reith Lectures, so it's not exactly new. Progress with marrying science and politics has certainly been slow but, for the sake of justice, it is important to try to stop bad things happening. If used wisely, science can predict, prevent, repair, improve and restore. It can be an alternative paradigm to that of the bare-knuckled model of consumptive economics. It is a paradigm which sees government as a mechanism principally for managing risks and, therefore, creating the circumstances within which human ingenuity can thrive of its own accord. Within this book, I want to show that this is a pragmatic approach to mixing science with politics. It is built on what I call *natural realism*,[27] which, in broad terms, concerns the acknowledgement that all we aspire to do is ultimately constrained by natural processes.

Part 1
A troubled marriage

A recurring concept through this book is that of *systems*. I have already referred to political systems, scientific advisory systems, administrative systems, government systems, warning systems and the collapse or bifurcation of systems. Later in the book, I pick up this theme in more formal terms because governments themselves expend much of their resources managing systems, and science has much to say about how to do this.

Our most basic understanding of systems is rooted in the laws of thermodynamics. They are normally formed of many interacting components contained within some sort of defined boundary. They are often interconnected with one another, or they may even be nested within each other.

Some systems are purely internalized to government or society, such as educational systems or the international financial system. Others are natural systems, such as ecosystems, some of which operate at large scales, for example the weather system or the solar system. Small-scale natural systems might include the systems of control inside individual living cells, something which attracts huge amounts of research interest from scientists, especially in cancer research. But at medium scales many systems are hybrids involving both societal and natural components. These socio-natural systems, such as the food system, involve strong interactions and dependencies, sometimes called 'feedbacks', between the societal and natural system components.

In simple terms, a lot of science is focussed on describing the *state* of systems at any point in time and how systems then change through time, something known as system *dynamics.* Technically, if we understand both state and dynamics, then we have useful knowledge of the system. Many researchers spend much of their time focussed on individual components of systems, but they are normally conscious of how those components connect to others, and, indeed, they may be trying to tease out the processes of cause and effect within systems. They often try to solve the problem of how to intervene in systems in order to shift their dynamics in specific beneficial ways.

If politics, therefore, is a system of collective decision-making, how should scientists respond in this case? An understandable response caused by the *scientifico-political predicament* has been to keep well clear, to stay outside the boundaries of that system and only to interact in very specific, prescribed ways. However, rather than having two, separate systems of thought and action – science and politics – I suggest what we really have, and need, is a single scientifico-political system. In this part of the book, I consider the current state of this system and provide some reflections on its structure, organization and historical dynamics.

1

Beyond two cultures

It is remarkable how often the *two cultures* discourse is referred to by scientists, even if not by name. Based on his experience as a scientist advising the UK government, Solly Zuckerman said that 'when scientific enquiry steps out of its own ground and treads into the area of value and moral judgements, it starts to become something more than science, and something which begins to partake of the controversial character of economics, or even politics'.[1] This was an acknowledgement that science was different from the cultures in normal circulation within society. Patrick Vallance, another British science adviser, also said it was important for scientists to 'stay in their tramlines' when talking about the docking between science and the processes which generate and deliver policies made by governments.[2]

I identify with these feelings because I saw for myself how the intervention of scientists could irritate politicians and it often made them less likely to listen to advisers like me or Vallance. 'Can't you have a word with these people?' was what I would sometimes be asked, in the vain expectation that science, like politics, was about 'whipping' colleagues to align with the tribal norms. Politicians understand that scientists are more trusted by the public than them[3] and therefore the scientific message needs to be managed. The 'tramlines' here are about keeping science out of politics – one approach to solving the *scientifico-political predicament* – and it is an extension of what has become known as the *two cultures* discourse.

This discourse was brought into sharp focus in 1959 when Charles Percy (C.P.) Snow – physicist, bureaucrat, novelist and politician – gave a now famous lecture at the Senate House in Cambridge entitled 'The Two Cultures'.[4] In it, he suggested that the pathology at the centre of government resulted from mandarins' or politicians' specific educational backgrounds. He saw an emergence of deeply embedded dualism within society[5] which is both structurally and culturally codified through educational systems. On one side are, broadly speaking, lawyers and graduates in the humanities; on the other side are mainly scientists, engineers and perhaps medics involved in research and education. Snow's proposition was that government is structured to exclude Apollonian reasoning and empiricism in favour of Dionysian idealism and aestheticism. He thought this meant governments lacked diversity and an essential leadership and decision-making capacity[6] and this pathology afflicted both the legislature and executive branches of governments. They were largely populated by a social caste which was trained to think and debate very differently about problem-solving than scientists. Of the annual intake of high-flying graduates for the UK civil service in 2020, about 90 per cent were arts and humanities graduates, with the remaining 10 per cent coming from science and engineering. But even this 10 per cent would become subject to coercive and irresistible re-training to conform to the civil service norms. One senior civil servant with whom I worked, and whom I saw as exemplary of the 'other' culture, once told me that her university degree was in physics. Her admission seemed laced with a feeling of relief that she finally had found somebody to whom she could impart her dark secret and who would understand the deep conflicts which she experienced. Put simply, the *two cultures* proposition really describes two norms: one is comfortable with ambiguity and subjective, manageable and pliable truth; and the other, which is in the minority, is uncomfortable with those positions and wants to expose and resolve ambiguity and to seek objective truth.

Certainly, some of those with whom I worked when in government (including senior government ministers and senior civil servants) had known each other at university, or when at school. A few were even related to each other, so if running the country was not exactly the family business, it was certainly the business of those from a specific tribe. Even though a counter-culture also exists which rejects this kind of elitism within politics, often represented by more socially leaning political classes who emphasize their 'grass-roots' origins, these people carry other tribal and cultural baggage. The cultures referred to by Snow are co-moulded sociologically, genealogically and education-ally and they apply just as much to those who occupy the legislatures of nations as to the national bureaucracies.[7] But their common trait is that they have a poor under-standing about how science, if used well, provides a central plank for the structuring and functioning of society. Some may even be actively hostile to scientific thinking, talk-ing down scientists and projecting them to the public as freakish boffins whose mad schemes would lead to disaster unless they were there to prevent this. The *two cultures* hypothesis originates from the frustration felt by scientists when their contributions are ignored, misused, misrepre-sented or under-appreciated by these dominant political classes.

A case can be made, I suppose, that this division is a result of a long history. Science is a relative newcomer to the business of politics and the machinery of government. Science has been slow to emerge from the shroud of hocus-pocus because at one time it was blasphemous to opine that there might be a more fundament truth than that promoted by priests. Proto-scientists had to take great care for their own safety, so early surreptitious efforts to build communities of thinkers could be indistinguishable from the esotericism of occult Rosicrucian-styled secret societies like the freemasons.[8] Such an example was the 'Invisible College' of the 17th century, which was the precursor of today's Royal Society, one of the world's pre-eminent scientific institutions. These roots meant that science was always intentionally kept distant from politics.

The Enlightenment came about because we started to systematically distinguish truth from hocus-pocus. Once that touch-paper was lit, then the explosive power of science drove rapid change, but apart from the challenge science presented to old hegemonies, it was what it did to us as people which worried many.[9] The rise of Romanticism and Aestheticism over 200 years ago was a response to the changes wrought on people as a result of the dehumanizing consequences of innovations made possible by science through the industrial revolution. They are carried through, even into present-day politics, as the two cultures, which arguably is partly driven by the fear of what science can do *to* us rather than *for* us.

The *two cultures* can also be seen as an extrapolation of mind–body dualism. This branch of philosophy, which was particularly prevalent in the early 20th century, articulated how humans can struggle to conceptualize a world beyond their own limits of perception and imagination. On the one side were the *positivists*, who brokered scientific method, and on the other were the *aestheticists* and social theorists.

The philosophy of language was especially important because it is the mechanism we use to express and communicate complex ideas. Snow, who was situated in the middle of this philosophical divide, sought to capture how these contrasting views of cultural change were being expressed within the critical governance structures of society.

In the early 20th century, the analytical philosophy of thinkers like Bertrand Russell and Ludwig Wittgenstein aimed at marrying how we express ideas of the mind in the language of logic and mathematics. In spite of their work and that of others, throughout much of the century the philosophy of language seemed to lose its way because it had the effect of emphasizing the primacy of mind over matter – interpreted as the internalized interests of people over the externalized effects of Nature – and the ultimate constraints that our own sensory and cognitive abilities place on our interpretation of the world around us. It followed the lead of the early 18th-century clergyman and philosopher Bishop George Berkeley, who saw physical reality as subsidiary to mental and spiritual reality. His was an entirely internalized philosophy but, by kicking a stone and showing that it moved and hurt his toe, Samuel Johnson objected to Berkeley's view, a view which also resonated with the age-old idea of the primacy of humankind over Nature, something which in turn had been a tenet of Christian teaching since Augustine of Hippo in the 5th century CE. Johnson's point was that, pragmatically, one would not come to such a bizarre conclusion that the stone was purely in the imagination. The stone was part of Nature and his toe interacted with it. Augustine recognized this as *natural law*, which we might see today as the *Laws of Nature*. In the absence of this understanding, we surely descend into solipsism.

In modern terms, this concept of mind over matter has interesting philosophical consequences concerning, for example, how we would prove that we are not living within a virtual world as a part of a computer simulation.[10] Certainly, our rapid transition during the COVID-19 pandemic to using 'virtual' technologies has accelerated this notion. There is a distinct possibility that in the foreseeable future people could spend most of their lives in virtual worlds, effectively living and even dying within a simulation.[11] In some senses, many of us are already in this state thanks to our inter-actions with chatbots and other virtual 'creatures'. Virtual people abound in our lives, mostly represented as 'celebri-ties', which are manufactured personas built upon a human frame, although there may be no reason why the human frame will be needed in future. However, we have moved on from Bishop Berkeley's view of the world because advances in instrumentation, experimentation and statistical analysis mean that we no longer rely on our own senses to inter-pret reality. As the biologist Richard Lewontin suggested, there is no escaping the tight co-evolutionary relationship between humans as scientists, society and our interpretation of reality.[12] It is not possible to consider any one of these components in isolation from any other. The important conclusion is that politics is intimately dependent on not just the reality of the world around the society it supports, but also the ways in which we develop our understanding of that reality, something which lies in the domain of science.

Unfortunately, none of these insights were clear through-out the 20th century, when most of those students taught in philosophy, many of whom would ascend the political and bureaucratic ladder, received only a partial and very biased education about the true nature of reality. Even then, many of the Laws of Nature were still relatively new, debated and not commonly known. (I frequently had to educate minis-ters on the folly of creating policies which tried to defy the laws of thermodynamics.) The primacy of our imaginations and our sociality over the world around us was locked into how the people who ran nations thought about the world; it was these kinds of attitudes which C.P. Snow observed among bureaucrats and politicians. Like the Romantics, they hankered for a future which looked like the past when nothing of the sort was likely. Inevitably, this will have had consequences for how regulations and laws were framed and implemented. It certainly created a false impression, still very prevalent even among academics (but, I suspect, almost universal among politicians and the public today), that the products of our imaginations are as true as is the scent of a flower or the setting of the sun.

The confusions about these things led philosophy into a parlous state in the late 20th century and stimulated the statement of exasperation from Stephen Hawking that 'philosophy is dead'.[13] His view was that philosophy had lost its way and had failed to keep up with advances in science and the rise of what is now, in philosophy, known as *scientific realism*. Mainstream philosophy had isolated itself from the developments in computer language inspired by Russell's and Wittgenstein's analytical philosophy, which has led to the blooming of an immense system of computer languages. This has begun to reveal emergent properties in artificial intelligence involving machines which can speak and think in similar ways to people, suggesting that the *mind* is not unique but instead is also an emergent property of networked computation. Moreover, the more we look in on Nature as Hawking did, the more we also see the characteristics of networked computation reflected back at us. Rather than defining machines as surrogates for people, this tends to define people as machines and they become component parts of much bigger machines, an insight which is equivalent in its social and cultural impact to the 19th-century revelation that defined humans as apes. This view of the world questions the very basis of *free will* as defined by Augustine.

At the most instrumental level, humans are simply networked agents who have a capacity to communicate using language at a rate of about 39 bits per second, which is only roughly one-millionth of the capacity of the computer on which I am typing this text. The result is that we are already cyber-physical entities utterly dependent on networked computation. The Romantics would have loved the new dimensions of imagined reality and the potential for creativity this would have opened for them, though they would have abhorred its implications for who they were. But more seriously, scientific understanding of the 'complex' dynamics which result in emergent properties of systems like the brain are beginning to bring rationalism to understanding values-based judgements which have been so slippery in the past, potentially overturning the *fact–value distinction* highlighted by Leo Strauss.

My personal awakening here was to see strong parallels between my own research on the economics of Nature and the economics of humans. The deeper I dug into the currencies used to create value, and the social contexts of complex non-human organisms operating in near-natural conditions, the more I saw that politics, which is so dominant within human society, is just an extension of what happens within Nature. Both crystallize in the end to networked computation.

Hawking was exaggerating. As the contemporary philosopher Peter Singer said in response, 'How could philosophy be dead when there is such a lively discussion of how we ought to live? Science alone cannot tell us the answer to that question.'[14] Science itself had a problem which it needed to address because the voice of science was not being pitched at the 'how we ought to live' problem, even if it could say a lot in answer to this question. Pursuit of the *two cultures* model of behaviour, including the pro-active separation of science from politics, had meant others were left to answer this. Instead, the 20th century saw science being established as the technical support function for society, and scientists generally seemed happy to fall into that role. They never attempted to be a part of the debate about 'how we ought to live'. That question had been ignored by scientists and captured by the philosophical aesthetes in important intellectual centres which educated many of the people who would dominate politics and public administration, the people sitting on the other side of Snow's cultural divide.

Snow's observation was not an endorsement of the situation as he found it. In a later lecture at Harvard when he expanded on the rationale for his description of two clashing cultures, he never proposed a solution.[15] Contemporaries within the 'British system'[16] like Solly Zuckerman and Peter Medawar[17] mainly reinforced the *two cultures* hypothesis. In his book *Beyond the Ivory Tower*, Zuckerman recognized the cultural divide as a problem for all the same reasons as Snow,[18] but he accepted the hegemony of the other culture (that of the political elite) as a *fait accompli*. In response, his solution was to create a fix which was the seed of the current system of scientific advice to government. It emerged in parallel with a very similar system arising in the US stimulated by Vannevar Bush,[19] who was science adviser to two Presidents, Franklin D. Roosevelt and Truman. Even if both Zuckerman and Bush saw importance in the rising power of science, however, neither acknowledged the need for science to function increasingly as a way of thinking about problem-solving within the political calculus rather than just as a post-political service for government itself, a kind of analytical donkey for the implementation of ideology.

Much subsequent reflective sociological analysis has also done very little to propose a different solution.[20] Indeed, the subservience of science to politics has been taken as the normal starting point for those analyses. Science functions as an adjunct of politics, or even to do the bidding of politics, rather than as an integral part of the political process.[21] In the view of many on both sides of the divide, never the twain shall meet except in highly managed, almost ritualized, interactions.

A few sociologists have seen science differently and suggested that the production of knowledge, which is mostly the job of science, is an increasingly collaborative and open process across society. Their view is that a lot of the old job demarcations resonant of the *two cultures* are disintegrating.[22] They cite the rise of large collaborative mission-based scientific projects, like the Human Genome Project or the search for COVID-19 vaccines, as evidence of alignment between social needs defined in the domain of politics and scientific endeavour. In addition, there is rising emphasis on 'traditional knowledge' as a source of wisdom.[23] These sociologists have a point, and a more egalitarian approach to knowledge creation is likely to deal with some deeply engrained biases in how science works and in what we know.[24] But great care is needed not to subjectify knowledge. Just because somebody claims they have knowledge does not mean that their knowledge is reliable. Reliable knowledge can only be generated under specific conditions, and mostly these conditions do not occur within the kind of collaborative environments which exist across society. Reliable knowledge can exist if it is generated heuristically (e.g. by 'tradition'), but this can make a much more limited contribution than it appears because the underlying reasons for knowing something are not normally understood.

Therefore, people who think that knowledge can be generated collaboratively across society need to specifically and deliberately define the constructs used to verify that that knowledge is reliable. The social amplification of falsehoods (caused in part by the revolution in social media) has emphasized the need for considerable care when imagining that useful knowledge can emerge from unsupervised and *ad hoc* social processes. The key challenge is how to judge the quality of knowledge emerging from embedded societal processes, and especially whether it has been manufactured with the intent of reinforcing ideology as opposed to the honest intent to reveal something true about the structure and function of Nature. Not everything that emerges from that mash is valid, and much of it just regurgitates hocus-pocus.

It is often difficult to tell the difference between serious knowledge and hocus-pocus, as is becoming evident with the emergence of AI large-language models. Falsehoods carrying the badge of authority are very dangerous and, as I shall explore later in the book, these are very prevalent in governments and across the whole of society. The rise of some types of artificial intelligence, which have no way of verifying the authenticity of the knowledge they use, is likely only to exacerbate these dangers. Indeed, knowledge generated using AI has many of the same problems as exist with socially derived traditional and heuristic knowledge.

One can start to see why the *two cultures* are sustained. When there is a need for a clearly disciplined way of generating knowledge so that it can be badged as reliable, then it is not surprising that a cadre of specialists arises who can generate reliable knowledge and provide appropriate authentication. However, the *two cultures* is really a problem of inclusiveness and diversity in politics, where the mixture of talents needs to be rich enough to include rationalist thinkers, authenticators, political idealists and pragmatic implementers in balance. The essence of the problem with politics today is that there is an imbalance. Science is often left on the sidelines trying to be heard and only being listened to when the political leadership thinks it needs help.

Perhaps history also explains a lot here too. Rory Stewart, former diplomat, teacher of classics and UK government minister – to whom I was an adviser – thinks of rhetoric as a pathway to truth.[25] Certainly, within the context of classical Greece, rhetoric was a highly valued way of building adversarial argument in a sort of dialogue which, if done well, can converge on a kind of philosophical truth. For example, Plato's dialogues come over as carefully choreographed intellectual fights. It is this kind of approach to debate which has led to our adversarial legal system and, also, our adversarial politics.

However, political argument ends up as sophistry if it is not grounded in a form of realism which aligns with Nature. Without this grounding, rhetoric is a very odd way of getting at truth and, at best, it can only settle on a specific kind of truth involving social consensus, which may bear no relation at all to the truths about Nature. It is the truth that Bishop Berkeley would have recognized. But we do not use rhetoric or social realism to agree mathematical truth such as the values of pi, Euler's number or the golden ratio, which recur through the structure and dynamics of Nature. We might debate whether information is destroyed by black holes, or the function of extra-cellular vesicles, but this is a debate informed by knowledge and is used to define the new questions which need to be addressed. It is also a debate informed by rules of procedure. It is not about asserting truth by winning an argument in whatever way comes to hand; it is about demonstrating truth through proof or probability. Rhetoric might be a pathway to a certain kind of socially accepted truth which places us firmly within the sociosphere or the social bubble of society, or within the political culture, but that does not mean that those truths are founded in the realities of Nature.

The problem created by sophistry is that it attracts people into professional politics whose ethics allow them to be comfortable with ignoring realism, and perhaps even with lying. There are no grounded reference points or standards inside the social bubble to expose their duplicitousness, and therefore duplicitousness thrives. Manufacturing social truth is a successful strategy because these people tend to rise to the top of adversarial political systems. But, as Plato recognized, just because one is good at rhetoric – mainly involving making promises which cannot be delivered but convincing people that the opposite is true – does not mean one has the skills needed to run a large complex business like government.[26] The least bad result is that we end up with incompetent political leaders[27] and bad government. The worse result is we end up with tyranny. Either way, it manufactures injustice.

Accepting the idea that there are *two cultures*, as appears normal in the present, is part of the pathology which pervades politics. It is prevalent because it is comfortable both for politicians and for scientists to 'stay in their tramlines' rather than to collaborate constructively and with mutual respect. Moving beyond *two cultures* requires those who promote its existence to stop and to work proactively to undermine its continuance. This applies equally to politicians and scientists.

2

The anatomy of a troubled marriage

At the height of the COVID-19 pandemic, an editorial in the *British Medical Journal* complained that 'When good science is suppressed by the medical-political complex, people die.'[1] The Nobel Prize-winning author Kazuo Ishiguro also recognized the critical role of science when in discussion with Venki Ramakrishnan (also a Nobel laureate, but for chemistry) he said,

> It seems to me that in this past year, we've reached the peak of two opposing ways of approaching truth. On the one hand, we've had a search for truth as it exists in your world, in the world of science, and we've come to rely on that desperately. All our hopes are placed on you being able to tell us what's happening, how we get out of this situation. And I think we really appreciate the fact that all the discussions you have are based on evidence and rigorous method. Contrasting to this, particularly around things like the US election, we seem to have this completely different approach to truth, which you could summarise by saying 'whatever you feel with sufficient conviction is the truth'. And the evidence is almost irrelevant. It's your emotions that give the truth validity.[2]

Certainly, when good science is not used appropriately, or even when bad science is used at all, people suffer. This makes the use of science essential to achieve justice in society but also a point of legitimate debate. When should we use it? How much should we use it? Who should use it? What does it tell us? In what ways and circumstances should it be used? Always? Never? Sometimes? How should we act based on scientific information? These are points of ethics; Aristotle pointed out that ethics is closely meshed with politics, and some think of Aristotle as the first scientist. Politics is the process of deliberation and debate used to explore these questions and to decide on policies which help to create collective action. Like the partners in a long marriage, both politics and science have become co-dependent and inseparable.

The partners in this marriage frequently suffer from a lack of mutual empathy. Politicians often want to ignore the inconvenient truths from science when they conflict with their socially manufactured versions of truth. Scientists become enraged about this because it is self-evident (at least to them) that this is reckless. It creates a chasm between people's expectations and their real experiences, as exemplified by the surprising arrival of COVID-19. The reality deficit in politics causes problems to roll in like relentless surf on a beach soaking up the intellectual and personal capacity of people in government assigned to deal with these problems. Scientists tend to emphasize the need to understand the real cause of the waves and to stop them at source. Politics, by contrast, is reduced to crisis management, attracting the kind of people to the political side of the marriage who can cope psychologically in a kind of madhouse. Co-dependent these partners may be, but they are from opposite ends of a psychological spectrum of risk perception.

Scientists have invented coping strategies to help make the marriage work. For example, a common view among scientists is that 'a policy decision is a complex political calculus of risks and benefits, and scientific evidence is just one input'.[3] I also used to believe this, but the core idea that 'scientific evidence is just one input' risks placing science on a par with religious and other ideological belief systems. Some of these would have us building decisions in the present around the body of knowledge which guided mediaeval governments, the same knowledge which reinforced inquisitions, witchcraft, alchemy, theocracies, heliocentrism or the primacy of humankind over Nature. These views are worryingly nearby, even in the modern age. At his General Audience in 1978, Pope John Paul II said, 'Justice is, in a certain way, greater than man, than the dimensions of his earthly life. Every man lives and dies with a certain sense of insatiability for justice, because the world is not capable of satisfying fully a being created in the image of God.' In his search for justice, John Paul II is replacing the primacy and reality of Nature – 'the dimensions of his earthy life' – as the arbiter of justice with that of humankind in the image of God. He is abstracting humans from the world, absolving them of responsibility for properly understanding Nature in order to achieve justice. It is this kind of odd exceptionalism which means we feel comfortable with the dangerous notion that science is 'just one input'.

In practice, however, governments are rarely as ideologically rooted as Popes, and they have a tradition of building public policy around trying to predict the consequences of actions, especially within Western-style democracies. Experience that science can lead to better outcomes means that science has increasingly ingratiated itself into policy decisions. But there is a tension between support for the role of science in predicting and assessing the effectiveness of policies and its role as an activity used to support strategic outcomes. The political side of the marriage would often prefer scientists to be distant from immediate problem-solving and the potential this has to undermine ideologically based policy. It should, in these circumstances, simply be there to provide technical support like an ambulance to be called after a car crash. The politician in the marriage would rather not have a partner who is continually sniping at his or her behaviours. Consequently, science constructed to mark the homework of politicians (also known as *science for policy*) attracts much less investment and attention from politicians than its effective opposite (*policy for science*) where politicians attempt to build a scientific endeavour to fit with their objectives. In this latter role, science is often at arm's length from the immediate political waves and is seen as a strategic investment to support the development and growth of nations but sometimes politicians are very directing in what science is done under the banner of policy for science. When it works at its best, knowledge itself and the institutions which support its creation are seen to be important parts of national infrastructure. In practice, this is usually focussed on magnifying *private* interests, by placing scientific inventions within markets which have the capacity to develop and scale up their impact.

There are also fundamental, cognitive differences between the partners in how they understand and interpret knowledge. If knowledge was in the form of a picture, most people (including politicians) see that as a positive image, like a picture of objects in a room. But scientists see it is a negative image, or an image full of gaps rather than objects, something equivalent to the Japanese concept of *ma* or negative space. This is because in science the concept of *uncertainty* is pivotal – the focus is on what we do not see in the picture rather than what we do see. And with uncertainty comes fallibility. Policies tend to be constructed around a set of 'knowns', but this creates problems – the waves hitting the beach – because these 'knowns' are usually miniscule compared with the magnitude of the 'unknowns' – the space between. When trying to construct policy, politicians call for 'evidence', which only provides the positives of the image and creates a highly biased description of knowledge. Policies are then set up as infallible propositions when they are nothing of the sort. Scientists, therefore, often end up as sceptics, pointing out all the unknowns, the gaps in knowledge and the points of failure in policies, something which those on the political side of the marriage just find irritating.

This is a fundamental difference in outlook which is illustrated by the difference between the 'shades of grey' view of scientists and the binary 'black and white' view of the world reflected in political and legal process. In politics and the law, propositions often have binary states like *right* or *wrong, compliant* or *not compliant, guilty* or *not guilty,* or *liable* or *not liable.* These legalistic propositions are built on a presumption of certainty rather than uncertainty; they are a product of adversarial rhetoric and are a positive rather than a negative image of reality. They contrast hugely with scientific argument involving the balancing of probabilities. Indeed, the greyness of science as a guide to decisions leads some people to the extraordinary conclusion that it is a poor basis for decision-making. When Oliver Houck, a Professor of Law, likened the relationship between science and the field of public policy to a 'troubled marriage',[4] he thought this was caused by the existence of scientific uncertainty. The problem, in Houck's view, lay on the scientific side of the marriage. But it has to be better to make decisions based on a realistic assessment of uncertainty rather than to ignore this and assume false precision? That cannot be a proper route to realizing justice. Amazingly, however, there are people, presumably like Houck, who think it is.

When I entered government, science was beginning to make progress with changing this. Policy professionals were starting to speak in terms of *proportionality* in decision-making, which was about colouring their decisions with conditionality based on the level of certainty which existed in the knowledge used to support the decision. Making decisions is unavoidable, because the world must move on, but it seems so much better to do this based on a proportional approach than to conclude, as Houck did, that science is a poor basis for decision-making. Indeed, precisely the opposite is true.

The law, and much of decision-making in government, is fundamentally subjective and values-based. As Snow would have observed, this is hardly surprising given that many of those people who run politics are from a legal background. I expect that those who practise law, some of whom end up in politics, imagine that the law has objectivity at its core, but 'show me the judge, and I'll tell you the law' is a powerful conclusion of one analysis,[5] and there is no better place to look for confirmation of this than the political gerrymandering around the composition of the US Supreme Court, which affects judgments on issues like abortion rights. The subjective reasoning within politics and the law does not deal well with uncertainty and fallibility. It is inconvenient in law and politics just to say 'sorry, we are not sure'; guilt or innocence needs to be established, blame needs to be apportioned, decisions need to be made and often, politically, the only way forward is to assert one way is right and another is wrong. Failing to realize that reality is not so black and white can result in some odd judgments.[6]

Signs of stress in the marriage came to a head on 22 April 2017 when in a coordinated action across about 600 locations around the world, scientists took to the streets to demonstrate discontent. This 'March for Science' was not about the usual reasons for these marches: low pay, social exclusion, anti-fascism, anti-socialism, anti-globalization or anti-establishment sentiment. It was simply that scientists were fed up with being ignored by the political system. Politicians like Donald Trump, who was President of the United States at that time, wanted science as a service to support their ideologies and the science community was saying 'no'. If you want one part of what we do, then the deal is that you get the whole lot and you need to use the whole lot.

The entrainment of science to ideological, or 'normal' politics has a long history: politicians through the ages have assumed that science is a service to them and their objectives. When the British Labour politician Harold Wilson said in 1963 at his annual party conference that the future Britain 'is going to be forged in the white heat of this [technological] revolution', he was probably right, but even then there were those who doubted the virtue of such a vision. And the prophetic words 'we go to the moon' delivered by John F. Kennedy at Rice Stadium in Houston, Texas, on 12 September 1962 reflected the same optimism of the times. These were signals from political establishments for scientists to lead positive change within society, but on their terms. Many scientists happily acquiesced and, in those days, I suspect few of them would have questioned whether those were appropriate goals.

But some did, and the month after Kennedy's speech, in October 1962, the world held its breath as the US and the Soviet Union squared off in a game of nuclear chicken concerning the deployment of nuclear missiles in Cuba and Turkey. The destructive capacity brought by scientific invention had, in the hands of the political classes, made the world a more dangerous place to live in than at any time in history.

These dangers were getting out of control. Only fifteen days after Kennedy made his speech, Rachel Carson's book *Silent Spring* was published.[7] This is now one of the most iconic descriptions of the devastating effects on the planet of technologies invented by humankind. Carson's careful analysis was like a slap in the face of the techno-dreamers and moon gazers. In war, destruction wrought by kinetic weaponry is sudden, focussed, brutal, identifiable and definable. In contrast, the chemical war which Carson described was being waged against Nature, and by implication against humanity too. It was diffuse, deceptive and pervasive. In the seventy-plus years since the book's publication, the chemical weaponry has become less overtly brutal but it is even more diffuse and cryptic. Much of this weaponry is discharged without understanding its true destructive power. The 'heroic age' of science reached its zenith in the first half of the 20th century, giving us a capacity to manipulate Nature but with an insight somewhat like that of a child playing with a hand grenade. It showed us that heroism is not the same as righteousness. Carson was saying that technology had a down side as well as an up side, but politics had a habit of only focussing on the up side, the good times within the marriage. When I re-read her book fifty-five years after its publication and around the same time as the scientists were marching in protest at being ignored, it had lost none of its freshness. In my role as the adviser to the UK government about the effects of these chemicals, I was left wondering if anything had really changed.

Around the same time, I attended the World Food Congress in Des Moines, Iowa. This proved to be a demonstration of why nothing had really changed. The Congress was constructed largely as a collective worship at the feet of Norman Borlaug, the Nobel Prize-winning agronomist who pioneered the so-called 'Green Revolution' which saw rapid increases in agricultural productivity as a result of the joint effects of agrichemicals and plant breeding. What I saw at the Congress was, however, a festival choreographed by large global corporates whose slogan 'We feed the world' was underpinned by aggressive anti-competitive business models. Borlaug's imagery had been purloined as iconography like that of Che Guevara or Lenin. In spite of the adulation, Borlaug's legacy is certainly mixed: there are some who believe that the 'Green Revolution' has been responsible for creating a lot of current poverty and starvation.[8] He probably genuinely thought he was doing good, but he got captured by big business, which forced the peasant farmers of the world to accept ways of farming which were bad for them and their environment. They got drawn into harvesting the richness of their soils mainly for the benefit of corporate Europe and America.

Bullying partners are not just present within the capitalist model. In the 1930s, the geneticist Trofim Lysenko placed the ideological doctrines of communism ahead of the truths about the structure and function of Nature. The emerging knowledge of genetics was seen at the time in the Soviet Union as bourgeois or fascist science. Lysenko, by contrast, promoted the doctrine of Lamarckian evolution, which involves the inheritance of acquired, rather than genetically coded, characteristics. This was something which had been clearly shown as unfounded by the 1920s, when modern genetics began to be shown to hold the key to both evolution and the underlying mechanism of selective breeding.[9] With the support of Stalin, Lysenko became head of the Lenin Academy of Agricultural Sciences, and those scientists who opposed his doctrine were either executed or outlawed, thus exacerbating the under-performance of Soviet agriculture. As with Borlaug, it would be difficult to account precisely for the damage done by Lysenko's ideology, but *The Atlantic* has called him the Soviet era's 'deadliest scientist',[10] which must be truly saying something. His unfounded ideas certainly resulted in enormous levels of suffering on a scale comparable with that of the effects of the Second World War.

When personalities and ideology completely subjugate science, and when science becomes just one component of political ideology, then people die.

It might be hoped that we have moved on, but, like my impressions of how little we have learned from Carson's teaching, as early as 2006, Vladimir Putin was constructing state control over science with a new 'Law on Science and Technology Policy' which placed the Russian National Academy under state control and allowed Putin to approve the 'elected' president of the academy, something which exceeded even the czarists and communists and 'which dismayed leading academicians' at the time.[11] Even Lysenko was being put through what appeared to be choreographed rehabilitation.[12]

This has all the characteristics of a marriage poisoned by the controlling political partner, whether from a tyrannical political regime or from tyrannical capitalist institutions. Political bullying means that the scientific mainstream has, understandably, a strong tendency to follow political norms. Let's face it: scientists who rely on public or private funding for their sustenance are very unlikely to stand up to those who regulate that funding. Carson was herself taking risks by publicly questioning the scientific mainstream, especially within government and commerce. This was near the height of McCarthyism in the US and President Eisenhower was told she was 'probably a communist'. Others dismissed her as an 'hysterical female'.[13] Borlaug's patronizing view of Carson as a 'gentle, great scientist and authoress' was offset by an attack on her for writing *Silent Spring*.[14] Rather than championing 'revolution', Borlaug had become 'establishment' and sold out his free will and his ethical judgement. Carson, on the other hand, was doing what all good scientists should do and was challenging the norms.

These experiences demonstrate how the establishment on both the right and left of the ideological spectrum had captured and subjugated science. In the present day, this is harder to see in places like the US, but it exists to an extent through the over-riding presence of the state in certain fields of science, and particularly in those with any strong relationship to policies like defence, health, social mobility and the environment. In places like China and Russia, there is no pretence at dissembling. The main instruments of state power and control of science are channelled through their respective National Academies, which are controlled by the political partner in the marriage.

Carson had sensed the dangers of normalizing science but was not alone in seeing huge problems ahead. In the early 1970s, the Club of Rome produced its global analysis *Limits to Growth*,[15] which was the first systematic assessment of how humanity could hit the limits of global resources and, importantly, the limits of global capacity to absorb pollution. This was the time of the blooming of the 'conservation movement' when environmental pressure groups like the World Wildlife Fund started to be formed and there were official acknowledgements of the need for environmental awareness. The Council of Europe sponsored European Conservation Year in 1970. It was also at this time that we began to collect many of the data sets which are used today to measure change in aspects of the natural world, such as trends in species abundances, and which set base lines about what the natural world was once like. A common myth is that we know what the natural world was like in its pristine state. It is possible to infer what it was like but in reality there is no solid empirical foundation further back than about 1970, which post-dates much of the gutting of the richness of nature described by Carson.

In 1975, the anthropologist Margaret Mead (who was then the President of the American Association for the Advancement of Science) convened a conference about the endangered atmosphere to highlight the emerging sciences of climate change and air quality. This, and the other awakenings going on at this time, became contentious and began to bite politically. John Sununu, who later became the White House Chief of Staff to President George H.W. Bush, branded Mead an authoritarian Malthusian.[16] He was attempting to politicize the messages from scientific research as anti-libertarian, when Mead was simply stating the direction in which the facts were pointing. This was a case of shooting the messenger rather than listening to the message, of relegating the message to 'just one input' and conveniently trying to ignore it. A growing number of scientists have been emboldened to speak out on these and other kinds of issues over the past half-century. The March for Science in 2017 demonstrated a boiling over of frustration about scientists being systematically ignored. It was a demonstration of 'getting political'. As Sir Brian Hoskins (one of the UK's leading climate scientists) said at the time of the March, science and scientific evidence 'cannot just be dismissed on a political whim',[17] a process which was making scientists 'professionally depressed'.[18]

However, fear of science is likely to be a powerful reason why politicians attempt to control it. This fear is less about the negative impacts of new inventions and more about the fear of challenge to political hegemonies. *Science for policy* sounds dangerously close to *science making policy*. In his farewell message in 1961, President Eisenhower said that 'public policy could itself become the captive of a scientific-technological elite'. Vannevar Bush, the first science adviser to the President, also remarked just before the US joined the Second World War, 'It is being realized with a thud that the world is probably going to be ruled by those who know how, in the fullest sense, to apply science.'[19] The accretion of power by scientists apparently made Richard Nixon 'seethe with anger', and as a result he abolished both the post of Presidential Science Adviser and his Council of

Scientific Advisers in 1973. He cut himself off from objective reasoning, which is a step more extreme than using science as 'just one input'. Science, especially in the US, was rising to 'establishment' status, and this meant it was in the cross-hairs of thin-skinned political beasts.[20] In a modern context, the globalized science-based corporates in information technology, pharmaceuticals and agrichemicals, which are at least as powerful as most developed nations, suggest that Eisenhower's fears could be partly right. Except it is not really 'a scientific-technological elite' which is in control but the business elites which own the patents and the intellectual property generated by scientists. Scientific discoveries have been excised from the moral frameworks used for their creation.

The power of science has therefore risen within the autocracy of capitalist institutions rather than the democratic institutions designed for public good. Scientists have surrendered control of their inventions – as Borlaug did – to those whose interests are the creation of private rather than public wealth. The aristocracy which governs science has caved in and become supplicants to the idea of science as 'just one input'. Within the marriage, only one partner has control of the bank account and the other partner has often been so eager to access its housekeeping money that it is prepared to be obsequious and supine, rather than frank, when dealing with 'political masters'. The tacit deal between the political and scientific establishments has been that in exchange for public funding, scientists look after their own interests – something known broadly in the UK as the Haldane Principle – but remain distant from politics. They leave politics to the politicians. It may be one solution to the *scientifico-political predicament* and is a deal which suits both, but it does nothing to ensure that the products of scientific endeavour are a part of the pathway to justice within public policy discourse.

Others who have also been critical of the role of science and scientists in technology development have thought of them as often irresponsible.[21] The point made by critics is that scientists must carry some of the responsibility for not showing more moral and ethical backbone and for not shouldering responsibility when technologies go wrong, often after these technologies have been transferred into the hands of those who exploit them mainly for private benefit. Much more could be done to manage these risks,[22] and scientists have an important part to play here, but mostly it is left to lawyers, accountants, historians, bankers, businesspeople, civil servants and politicians to manage the emergent moral hazards.

This leaves some sociologists comparing science with the metaphorical character of a *golem*, which is a mostly gentle, generous beast from Jewish mythology but which can also be unpredictably dangerous.[23] Fear of the *golem* was illustrated by a report produced in 2013 by the European Environment Agency titled *Late Lessons from Early Warnings*.[24] It detailed a range of case studies of technology gone wrong, but as with all these kinds of analyses, it is a view laced up with all the benefits of hindsight. In reality, the *golem* is unpredictable, but this is partly because scientists themselves have ceded control of what they produce to an ethically blind market.

Carson's analysis was really saying that some parts of the scientific endeavour had lost their way and that under the domineering presence of politics it was no longer a force for good. Perhaps September 1962 was the time when the curve on the graph describing the rising progress of the utility of science in society dipped below the rising curve of the cost of science to society. Science was no longer delivering net benefit – not at least when its priorities were being set tactically by politicians. My reflections fifty-five years after the publication of *Silent Spring* told me that scientists had failed to become their own political force to look after the moral code guiding how the products of their ingenuity were used. Even today, the consequences of scientific invention go largely unchallenged by scientists. 'Over time, digital technologies have become more sophisticated, and now there is a massive, churning, finely tuned digital misinformation machine that has seized social media to ensure that a portion of the population doesn't accept science.' So said Holden Thorpe, the Editor of *Science* magazine.[25] I might go further by suggesting that social media is a new form of censorship of the kind which existed in pre-Enlightenment times when religious orthodoxies drove free thinking underground. Social media may be the product of science itself, and much loved by those who believe it is a route for the spread of liberal values, but it has been appropriated by politics and promotes conspiracy theories, and 'whatever you feel with sufficient conviction is the truth'. It is the antithesis of science.[26] The jury is surely still out on whether it is socially progressive or regressive. Social media is a case of scientists seeing their own inventions being turned against them without appreciating the irony of this situation.

Should scientists really be prepared to release their discoveries into the world and then to retire from the values-based chorus which ensues as the negative consequences start to appear? Should they be surprised if their own inventions are turned against them? It is important they take responsibility for what they produce and steward it through the moral hazards within society. There are sobering stories of scientists experiencing that sudden moment of revelation when the penny drops about the devastating consequences which could arise from what they have just discovered.[27] Their first reaction is never to tell anybody, never to let the *golem* loose, and to try to bury the terrible truth because, once it is released, they will have lost control of their own invention. The connection between genes and behavioural traits is one issue which could produce toxic outcomes (mainly associated with fuelling racism); an authoritative view is that research of this kind should not be conducted at all.[28] Perhaps if science had taken a similar values-based approach to its work over the past century, then the world would be in a much better place than it is now. Too often, I meet scientists who have never thought deeply about the ethics of what they are doing or they think that all research is by default a universal good. As a result, scientists have left the debate about values to others and ploughed on regardless. Robert Oppenheimer clearly felt that this ploughing on regardless was toxic because, following his leadership of the Manhattan Project, he spent much of the rest of his career trying to put the nuclear genie back in its bottle. The same mistakes are currently being made with artificial intelligence.

The March for Science was coordinated to coincide with Earth Day to train a light on how scientific endeavour is less about making more things for people to use and to destroy what they have around them and more about how we live. Science is becoming the warning beacon for humanity's inhumanity to itself and the planet. The March for Science was an objection to science being railroaded into being the crutch for the resulting rapacious exploitation of people and the resources of the planet. It was a cry from those who were, like factory slaves, being silenced, told to 'shut up and calculate!',[29] and urged to up the tempo of production even if what was being produced was being turned into a harbinger of doom.

Tasking scientists to keep producing things people want – from cures for diseases to feeding the growing population of the planet – is all very well and good, but this tasking involves a deal. When scientists warn about the effects of emitting greenhouse gases into the atmosphere, and other limits which humankind should not breach, the deal is that what they say is heeded. They took to the streets on the March for Science because they thought the other partner in the marriage was not complying with the norms of marriage.

Scientists are finding that the problems they define, however great they may be, do not compel politics to move. Holden Thorpe has also been more direct: 'Since the end of World War II, scientists have clung to the idea that if they stay objective and state the science, then the rest of the world will follow. As climate change rages and the pandemic cycles on, it's time to face the fact that this old notion is naïve.'[30] Just stating the facts does not compel solutions when those making the statement have no political power and when those with political power are prepared to mangle what science says.

3

Inside the politics factory

What is it like for somebody from a scientific culture to be dropped into the operational domain of the political culture, such as in government? In this case, 'government' is understood to be the place where policies are manufactured rather than delivered. It is where politics happens in its rawest form, where it becomes a Schumpeterian arm-wrestle for power and also where it becomes pathological. Why should I as a scientist have anything to do with this kind of politics? Like many scientists, I adhere to Peter Medawar's view that 'Nature will be the more beautiful for being brightly lit,'[1] but also that by lighting up Nature we resolve many of our own dilemmas reflecting on our own existence. Many scientists therefore simply turn their backs on the small-mindedness of politics, and understandably so. This is my story about why I did not.

My scientific interests had started from a very young age. My father was an ecologist working in the public sector and one of the pioneers of conservation science in the UK. Many of Scotland's greatest nature reserves were set up either by him or with his direct involvement. I was exposed, therefore, early on to much about the natural world and about the politics of how it interacted with the world of people. I moved quickly from being awed by the complexity, structure and beauty of Nature to being more interested in what made the natural world tick, reflecting a preference for understanding function over form. I progressed through the academic system, and after a spell trying to work out how seasonal cycles of food and light interacted to control the reproduction of wild mammals, including seals, rabbits and bats, I turned to using animals like them as windows into the dynamics of oceanic ecosystems. I wanted to reverse the normal logic flow involving asking how animals use signals from their environment into asking how I could use these changes observed in animals to tell me what is going on in the environment itself. I wanted to use animals as the large-scale equivalent of the canary in the cage once used by miners to detect the presence of poisonous gas. My 'canary' was, the delightfully feisty Antarctic fur seal, although I also studied southern elephant seals, penguins, albatrosses and whales.

These were the predators sitting at the top of the oceanic food chains of the Southern Ocean which forms a ring around Antarctica and they allowed me to look inside that large, complex ecological system to find out what was going on. They lived off the productivity of the ocean and exploited it across its whole surface and down to its greatest depths. Theoretically, they should respond to changes in the flow of energy through the oceanic food chains. They could tell me stories about the effects of climate change and the exploitation of resources by humankind. All I had to do was to learn their language expressed as a frequency-based signal within the rhythms and flows of their lives. I was using them in the same way as an astronomer uses a spectrometer to understand the chemical structures of planets and stars, except I was focussing my instrument on the oceans. I also reasoned that these animals had found a way of being 'sustainable' within their environment and that they may have been using some basic rules to achieve this. If we knew what these were and followed them our-selves, then we might also come up with a formula for a sustainable existence.

Fifteen years of work followed and boiled down to a single line on a graph. It showed that as the food supply for these animals increased, their performance in terms of their health and welfare increased rapidly at relatively low food abundance, but that once the abundance of food reached about one-third of the long-term maximum, their performance levelled off. There were two important lessons from this. The first was that in animals which had a long history of existing and surviving within this natural system, there were internal mechanisms which meant they did not exploit their own resources to the maximum extent all of the time. Indeed, they appeared to exploit only about one-third of the ocean's potential long-term food productivity. This apparent abstemiousness appeared to be their way of cushioning themselves against what Henry Kissinger called the 'inevitability of tragedy' within a complex dynamic system.[2] The second lesson was that when food abundance was in surplus for these animals, then it was probably alright for humans to step in and take some of this without fear of causing hardship to these other fellow creatures. This 'one-third' rule was probably a good guide, therefore, to the regulation of human exploitation of the ocean.

Another interesting result was that the currency these animals live by is energy. The volume of energy flow through the food web of the Southern Ocean was what set the level of their food supply. Lots of natural processes caused this flow to vary, sometimes by a factor of ten or more between years; when it failed and took the food levels below the threshold for their comfortable existence, this brought dis-aster for those animals which relied on it. These disasters were short-lived but were effectively like the short-term effects of climate change as ocean currents shifted for short periods before resuming their normal course again. When this happened, the beaches of the fur seal breeding grounds were piled high by the waves and tide with hundreds of thousands of pathetic, rotting carcasses of dead seal pups dumped by their desperate and starving mothers.

Together with colleagues working in other ocean regions, I was able to show that the same 'one-third' rule worked on a global scale. Animals which exploited ocean ecosystems sustainably never took more than about one-third of the productivity. This was not, of course, a choice on their part. Unlike us, they do not have collective decision-making mechanisms built into governance processes to come up with policies which help all these animals to abide by the rules. The rule was instead built into their life-histories and physiology by an evolutionary process which had selected for the most successful strategy. Compare this with how we humans have prosecuted fisheries where we regularly take 90 per cent of the available fish. Rarely do we ever think that it would be most healthy for us if two-thirds of the natural living resources of the planet was left untouched.

It was questions like this stemming from my own studies of how economics works in Nature which brought me to consider the problem of how to manage the interaction between people and the environment. This highlighted big questions about whether the global food system which has evolved to sustain a flow of energy and nutrients to over 7 billion people on the planet was really stable and sustainable. It was only a small jump from looking at the energy flows through the Southern Ocean which sustained a community of animals with very similar physiology and life-histories to humans to studying the energy flows through the global food system supporting people. While humans have built a much more complicated and technologically driven way of exploiting the natural productivity of the planet, it is not sufficient to just assert that this protects us from the same kind of effects grounded in the Laws of Nature which govern the lives of seals, whales and penguins. We need to show that this is the case, and much of my experience has suggested that we neither have a grip on this nor understand the many signals indicating that the system we have created for feeding the people of the planet has similar vulnerabilities to the system which feeds these predators in the Southern Ocean. We do not constrain exploitation of the natural resources of the planet with a view to the needs of future generations, and our food system only needs to fail once and life as we know it will cease. Instead of dead seals piled up, it will be dead people.

Studying Nature, 'red in tooth and claw', gives one a certain capacity for dispassionate analysis. Bringing this kind of analysis to discussions about how to create a better future by building the resulting insights into public policies has the advantage that one does not shy away from the unpleasant or the unthinkable. Neither emotion nor self-pity cuts any ice when it comes to how Nature deals with people. They are no different to those abandoned, starving seal pups crying for lack of food, wailing for their absent mothers with a shivering similarity to that of human babies. Over the many years spent in the field, mostly on remote windswept islands, I grew hardened, but the experience taught me the merciless brutality of Nature and I could not see how people could be immune to the same kind of fate.

Eventually, the wish I had to apply insights from studying natural process brought me into the role of Chief Scientific Adviser in the UK government on food and environment. This was based in the department responsible, the Department for Environment, Food and Rural Affairs – Defra – but I found myself in other cross-government roles such as one relating to earth observation using satellites.

In addition to wanting to make a difference, which many people could claim, I also wanted to continue my intellectual journey. I had discovered that what interested me were complex processes, and there was nothing more complex than the *politics factory*.

The word 'complex' here, as I shall explain later, is not a simple description of something too big for us to understand. Instead, it has a formal mathematical meaning. It is a description of the changes which happen as a result of many actors interacting with each other, often in a network. In chapter 1, I referred to network computation and, in this sense, I was beginning to see Nature in the same terms as a computer processing information. When I was studying the seasonal cycles of hormones, I was dealing with the interaction between tissues and cells, where communication of information was via chemical intermediaries. When I turned to large-scale ecosystems, I was dealing with the same concept, except that individuals and species were the networked agents connected mainly via energy flows. In government, I saw a similar structure where the networks were socially mediated and driven by a set of flows, which included money, power and agency. I was fascinated to learn more about how this worked and what could be done to make it work more smoothly, to achieve increasingly ethical outcomes.

Early on, I conceived 'government', and the wider diaspora of people, factions and institutions involved in the debates about how to live, as something which manufactured what we call *politics*. It seemed to me that there was a choice to make about whether one entered this factory, but that there were a great many people already inside in one way or another. Factions took on the character of institutions, like trade unions or industry representative bodies or other non-governmental organizations (NGOs) representing different views. One could see this as the ferment of democracy, but a question which hung in my mind when witnessing this was whether all these components of infrastructure within the factory, which mostly acted in their self-interest, added up to a process of decision-making which was in the collective interest. Was it in fact the tyranny of the masses where competition led to a Pareto distribution of power and interests – one where the great majority have little power and a small minority held most of the power? Was it anything but what most people thought it was – representative democracy? Was it, moreover, a case that the component of the politics factory which carried the label of 'government', which was acting in the public interest, was being burdened by the rest of the actors in this ferment with all the risks? After all, had the banks not fallen back on government to support them in 2008, and had nations not relied on government to haul them up by the bootstraps during the COVID-19 pandemic?

Even if government had a central role in the organization of society, it seemed to me that it was failing to share its risks among the other actors within the politics factory. Instead, the modus of behaviour was to point the finger of blame at government as soon as anything went wrong as a way of minimizing the chances that the finger might have to point back at themselves. Private interests in the politics factory often appeared to be insuring themselves by shifting their own burden of risk into a public liability. This happened at all scales, from people failing to take care of themselves and becoming a burden on the state to massive banks which fail and need state support to survive. Public institutions, such as healthcare services, are there specifically to deal with these kinds of consequences, but I saw politicians being too ready to accept this transfer of liability without really understanding how the total portfolio of liability carried by government was rising.

This business about risk-sharing was a battle-line exacerbated by the tendency for the actors within the factory
to inflate trivialities. Far from being an efficient mechanism for resolving problems, the politics factory appeared
to manufacture and amplify self-generated problems. It
created, rather than resolved, ambiguity about where risks
lay and who was responsible for insuring for them. I saw
the lack of tactical and strategic risk-sharing as one of the
fundamental drivers of toxic politics. It underpinned the
making of promises which could not be kept, thus creating undeliverable expectations and resulting in a need to
defend the indefensible meaning that everybody within the
politics factory was obliged to be devious and play a game of
deception rather than be honest.

I wanted to know how the politics factory worked. The
more infrastructure and actors there are within the factory,
then the greater the volume of politics which is manufactured. Most people step into the politics factory by impulse
or because of conviction or out of self-interest, but, to
remain sceptical, objective and ethical, it is important to be
able to step back out, to inspect it as a detached observer,
to be an *impartial spectator*, in the words of Adam Smith.
Some people enter and never leave and are never sufficiently aware to appreciate what has happened to them.
Many people who enter the politics factory are dishonest
with themselves and others about their condition and feign
indifference. Others are born into it and know nothing
else. I entered it by choice while trying to remain acutely
aware of my condition, although I saw myself more as a
subversive mole than a true player in the factory. I saw the
factory as a horrendous mash of self-interest which needed
to be challenged. For a scientist, the question of whether to
enter is an expression of the *scientifico-political predicament*, and scientists who are unaware of the predicament
are either already deeply embedded in the politics factory

or supremely susceptible to becoming entangled in its web.

My focus was on the 'government' component and, in my mind, I stripped that machinery back to its basics. By cutting through the complexity to understand the basic components of machinery, I grew to realize that 'government' was a lot simpler than it sometimes appears. It does three basic things: collecting *tax* to pay for its activities; *regulating* to ensure people stick to rules; and providing *support* by spending its money to address market failures and to provide services to people. Politics focussed on government is largely about deciding how to apply these three levers of control, but out of this simplicity, complexity can emerge. Indeed, the politics factory has a lot in common with a very large living organism, or a supercharged ant colony, because it has what is known as 'emergent behaviour'. All the ants, or people in this case, are programmed to do certain things and to respond in stereotyped ways but their collective behaviour – the growth and dynamics of the colony – is not something one could predict from observing any single component. Nothing controls it. It is purposeless but it rolls on regardless. Nobody has to understand it for it to function. It is much the same for all living organisms. The result, in the case of the politics factory, is that the emergent behaviour of the whole system is like that of a beast, and this means it can be unpredictable, even absurd. Let's look at an example of this absurdity, largely caused by one faction within the politics factory accruing disproportionate power to serve its own interests.

It was a Saturday in late September 2012 when a meeting request appeared on my phone. This was scheduled for seven o'clock on the following evening in London. I was in Scotland and managed to get a flight to London City Airport about four that Sunday afternoon. With luck I thought I would make it to Westminster on time.

The flight was slightly late but I arrived in the office of the Secretary of State for Environment, Food and Rural Affairs, Owen Paterson, at about five past seven. Other people were there, too, including the Cabinet Secretary and the Prime Minister's Private Secretary (who one day would ascend himself to become Cabinet Secretary, the nation's most senior civil servant). This was an ominous sign because these people from Downing Street never normally appeared in other departments unless there was a serious problem and a need for the Prime Minister to bring discipline to bear on an unruly faction or to sort out a political mess.

No indication had been given of what the meeting was about, but the other people around the table gave me a clue: the President of the National Farmers' Union, the most senior civil servant responsible for farming and food production, the Permanent Secretary of the Department (equivalent of a Chief Executive) and a few others. The Minister of State responsible for animal health and farming was on the speakerphone from his constituency.

The problem was that there was an epidemic of tuberculosis (TB) among cattle in the English countryside and this had been raging for more than twenty years. The politicians had made a rash promise to sort it out by killing badgers – animals that were politically, even if not epidemiologically, at the centre of a row within the politics factory. Claire Craig, the former Director for Science and Policy at the Royal Society, once explained that 'the long-running disputes about whether selectively to cull badgers to help manage the risks from bovine TB exemplify the reality of policy debates' tendency to be framed around ... charismatic entities and relatively simple narratives and choices'.[3] It is this apparent need for simplification and mis-specification of *complex* processes which leads to so many absurdities within the politics factory.

The question that evening in September 2012 was whether to issue farmers with licences to kill badgers in two regions of the English countryside. Technically, I stood in the way of this happening. For practical reasons, the population size of badgers in these two regions made a huge difference to the viability of the cull and I had to decide between two wildly different estimates of the number of badgers which had to be killed. The lower estimate had been produced by farmers themselves whereas the higher estimate had been produced by scientists. (In the end, neither estimate turned out to be particularly accurate.) It was not a surprise that the farmers had made their own job easier by coming up with the lower number. The high number, which I thought was more defensible, made the cull unviable that year so my decision to support it led to an embarrassing climb-down by the politicians.

The story of badger culling and the management of bovine TB was to occupy me throughout my tenure and is much too detailed to relate here, but it became an example of a problem which had been mis-specified within the politics factory and it serves as an excellent example of what might have been very different had science been better represented in the decision-making process.

TB is a devilish disease to treat. Unlike COVID-19, it is slow moving, deeply cryptic, irruptive and highly resistant to treatments like vaccines.[4] The hard fact, often overlooked in the politics factory, is that in the English countryside it was a disease of cattle, but the point which most people studiously ignored was that cattle are owned, traded and moved by farmers. In this sense, TB was a social disease, like a sexually transmitted disease is a social disease, spread as a result of human behavioural choices.[5] Rather than seeing it through the lens of farmers spreading the disease as a result of their choices, the politics factory focussed on other vectors of disease and the spotlight fell on badgers. Several lines of research had suggested that badgers were involved but were a minor component compared with spread caused by farmers themselves.[6]

Badgers became convenient scapegoats because neither farmers' leaders nor government ministers wanted to face up to forcing farmers to change their practices. Through the activities of their union, farmers had gained dispropor- tionate power within the politics factory. But the resulting absurdity is much deeper than the rather minor issue of the killing of badgers. Not only were many cattle farmers kept in business because of public subsidy, they were incen- tivized to tolerate TB by being compensated when their cattle contracted the disease. It is, admittedly, a devastating experience for most individual farmers to experience what happens when their cattle become infected with TB because most of them are killed, but the injustice felt by them is the result of the perversity of attempting to sustain a style of farming in the face of unmitigable risk. Government subsi- dies simply encouraged them down a pathway to injustice. They were trying to do something as sensible as growing tropical trees in a place with hard winters. By being wrongly incentivized, the cattle farmers were victims of the absurd- ity of the politics factory just as much as badgers were.

This problem is further compounded by an atavistic view within the politics factory that all this subsidy is needed to sustain the security of food supply. What is under- represented in these debates is the scientifically derived conclusion that most of what is produced by cattle – mainly meat and dairy – is over-consumed and produced with very low efficiency and extraordinarily high environmen- tal impact. On the whole, people do not need to eat food produced by these farmers. The tactical absurdity of killing badgers to control bovine TB was therefore compounded by a strategic absurdity associated with sustaining otiose farm production and an absurdly inefficient food system.

All this was also patently obviously unfair at a societal level. Why should people who self-designated as farmers be more worthy of income support than others like plumbers, electricians or care workers? The politics factory had concocted a way of using public money in a manner which was manifestly unfair to the wider community to encourage rather than prevent the transmission of a dangerous infectious disease, and to distract attention from this by killing animals which many people valued and many of which were not infected by TB at all. It would be hard to make up such a ludicrous situation.

The skew this created resulted in staggering injustices. While dealing with killing badgers, government was also supposed to be dealing with the task of improving air quality. I thought it was deeply ironic that while government was occupying itself killing about 27,000 cattle every year because they were infected with TB, plus similar numbers of badgers, it was prepared to overlook the killing of about the same number of people by poor air quality. Politics is about making choices, and in this case ministers had chosen to focus their time and resources on a dog's breakfast policy as a displacement activity to avoid saving people's lives. They were prompted largely by their interpretation of the wishes of their voters, many of whom hated the idea of managing air quality if it meant having to face the accusation that it was they, through the emissions from the vehicles they drove, who were really killing their fellow citizens.[7]

Intentionally killing badgers and cattle was much more remote and ethically easier to manage than admitting to intentionally killing people, so the latter was studiously ignored while the former became the focus of activity. In these circumstances, the wheels of the machinery within the politics factory are set to spin – to look busy – doing something relatively irrelevant. Bovine TB soaked up the intellectual, spiritual and financial resources of a section of government which should have been focussing them on something much more important. Who can possibly sustain an argument that science is values-free when it reveals such injustice?

The lesson is, of course, that the expression of aggregate self-interest in a UK-style democracy may not result in a generally improved pay-off. Everybody is worse off as a result of failure of government to do its job as a unifying core at the centre of the politics factory and to set the wheels of the machinery of government spinning in a meaningful way.

Such was the absurdity of the politics factory, there were times when I thought I was in a *Monty Python* sketch and could hear 'The Liberty Bell' march echoing around the corridors of government. When the UK decided it wanted to leave the European Union, and I asked if we had any idea where our food came from because I thought it was important to not provoke some sort of failure in food supplies, it evoked shrugs from civil servants and disinterest from politicians. The self-inflicted wounds generated through lack of insight into underlying processes and risks is truly breath-taking. Few of those who thought it would be a good idea to leave the European Union thought of the dangers this might create – it was truly the tyranny of the majority, and a very small majority at that. There was a tacit assumption that people, somewhere, somehow, who exist within the 'government' component of the politics factory will pick up the consequences and manage them so effectively that nobody else need care. Those people, who create risk without having to care about the consequences, have no appreciation of just how close to the edge of chaos we place ourselves through such cavalier behaviour. In exasperation, one of the country's most senior civil servants would mutter to me at moments of incredulous absurdity, 'You couldn't make this up could you?' Acting the part of Smith's *impartial spectator* means that looking inside the politics factory can become an almost surreal experience, equivalent to watching some sort of work of fiction, except that it is very real. The problem is that most people within the factory do not distinguish between fantasy and reality, a trait which I suggest is humanity's greatest weakness.

4

Rationalizing the politics factory

Anybody looking objectively and analytically at most government policies (of which badger culling is just one) through the eyes of the *impartial spectator* would conclude that they were mostly banal, a displacement activity masking the really important issues. But one of the most galling features of my experience of badger culling, and indeed many other issues, was that the application of science within the politics factory was of little help when it came to solving the problem. This happened because most of the research done was commissioned by people who wanted to solve the immediate political problem rather than the fundamental problem. Science then became the slave to politics, rather than the mechanism used to re-set the political agenda.

Science, in these circumstances, had become the victim in the *scientifico-political predicament*. Participating in politics without loss of objectivity is not something which, procedurally speaking, has been resolved, but there are two broadly opposing positions about how one might find resolution. One of these is summed up as the political component of the dilemma by the anthropologist David Graeber, who asked, 'What else are we ultimately except the sum of the relationships we have with others?'[1] Opposing this is the *scientific* component, summed up by the oft-quoted sentiment of Lord Kelvin:

> When you can measure what you are speaking about, and express it in numbers, you know something about it; but when you cannot measure it, when you cannot express it in numbers, your knowledge is of a meagre and unsatisfactory kind: it may be the beginning of knowledge, but you have scarcely, in your thoughts, advanced to the stage of *science*, whatever the matter may be.[2]

In Graeber's view, social relationships create meaning, and in Lord Kelvin's view, it is measurement that brings meaning. Both have merit and are correct in their own ways, so we need to work deliberately to balance the rational and objective, in the idiom presented by Kelvin, with the aesthetic, relational and subjective, in the idiom presented by Graeber. It is these contrasts which we need to balance within the politics factory.

Imagine drawing a horizontal line on a page as the axis of ideology on a graph. This might also represent Graeber's view of the world made up of social relationships. At one end is the label 'right' and at the other 'left', representing the *right* and *left* of the political spectrum of ideologies. Imagine also that almost all of what I would call *normal* politics happens along this axis.[3] There might be other ideological axes on a plane extending down through the paper, but we will keep this simple and just imagine that a single axis exists, an axis representing political ideology.

Now, imagine a vertical line drawn on the paper coming from the mid-point on the horizontal line with the label 'idealism' at the bottom of the line where it originates from the axis of ideology. At the opposite end is the label 'realism', something philosophers call *scientific* realism,[4] which

I see as distinct from *political realism. Political realism* is sometimes referred to as realpolitik. It is a form of realism which people like Henry Kissinger turned into an art form. It is the art of the possible within politics, an acknowledgement of proximate, pragmatic constraints about what can be done to manage problems. *Scientific realism*, on the other hand, while equally pragmatic, is often distal to those same problems. It is the root cause; an expression of the kind of constraints we see in, for example, the planetary boundaries[5] which define the safe operating space for humanity. As the axis of realism ascends from the axis of idealism, it gets closer to objective reality, to root causes. It is the axis to which Lord Kelvin is referring as something which can be measured. To paraphrase the philosopher Hilary Putnam, what makes a theory true or false, and which therefore allows us to observe reality as opposed to something imagined or abstract, is measurement external to us, that is, something that is not what we sense personally or is embodied purely in the structure of our minds or language. It is this kind of measurement which defines what exists up the axis of realism. It is everything from the measured balance of forces in a Cold War stand-off to measurements of the excessive use of nitrogen on crops and its impact on the planet.

Together, these lines represent a plane – the surface of the page – on which realism is orthogonal to ideology. This is a model of how to imagine the interaction of realism and idealism, and it is hopefully a useful way of depicting how science and politics interact within what I call the politics factory. Indeed, it is one possible highly simplified depiction of the politics factory itself. Very few issues will lie precisely on the line of ideology at the extreme bottom of the vertical line, but many people will attempt to place them there. Let's face it, a lot of the stuff we hear from right-wing libertarians, left-wing socialists or middle-of-the-road liberals sits firmly on this line. Much of it is imagined and does not refer to reality. Immigration policy may be informed by ideological positioning but it will also be pulled up off the axis of pure ideology by the reality of the mass movement of people and knowledge about what drives migration, such as population demography and wealth distribution among nations. Thus issues are pulled off the line to a higher level along the axis of realism by knowledge about how they relate to the real world.

People like me, scientists and many other free-thinking people, place particular value in getting issues as far up the vertical axis as possible. This does not negate the need for values-based decisions as they are represented by the horizontal axis, but it often matters little, at least to me, where those issues lie along that axis so long as they remain within ethically acceptable boundaries. I tried to ensure that my advice to politicians would be the same irrespective of their political view, or where they sat on the axis of ideology. I have advised politicians who came from opposite ends of the ideological spectrum and found I could easily work with both so long as I had this kind of model in my head. When asked by people who sometimes found this difficult to understand, I would simply say, 'My answer will be the same whoever asks the question.'

But there are important lessons for scientists, whose focus might be on *scientific realism*, from the experiences of applying *political realism*. These experiences turned Henry Kissinger into a much-derided character, some would say an amoral opportunist. It made him an enemy of almost anybody who saw politics through an idealistic lens. Although he could work with politicians of any persuasion, this meant his critics would say he stood for nothing and in this state one can easily lose one's moral compass. In the view of Kissinger's critics, his was a soulless method that appeared callous in its execution. Forcing compromise between conflicting idealistic factions arguably makes you a friend of nobody. One of the greatest challenges for free-thinking scientists interacting with politics is, therefore, to be taken seriously. When one is calling a spade a spade but when everybody else is imagining the spade as something much more exotic, then the tendency is for you to be seen as out of tune with social reality. But a big difference between the *political realism* of people like Kissinger and the realism of scientific discourse is that the constraints of Nature are absolute and unyielding and therefore they are ideologically agnostic. *Political realism* without the presence of *scientific realism* is rudderless and simply flexes to the prevalent ideology; this is probably what makes it amoral. It made Kissinger seem like an opportunist, bending his interpretation of reality to fit any political end.

Other professions also take the view that they are independent of ideology. Jurisprudence, for example, is technically a search for truth in any particular circumstance and should be agnostic in most circumstances to the ideologies of those involved. However, one important difference between the law and science is that in science there is normally feedback about how truthful a theory or judgement might have been – whether this resulted in a weather forecast or a medical intervention, the practitioner can get feedback about the truthfulness of his or her conjectures based on objectively measured outcomes. This leads to progressive, incremental and iterative improvement both in the methods used to draw conclusions and in the truthfulness of conjectures. In the law, judgments are made on evidence and, while sometimes new evidence is brought to bear after a judgment has been made and processes of appeal exhausted, in general, judgments are not subject to iterative feedback about their truthfulness. When feedback does happen, through the demonstration of injustice, it sometimes results in a clunky revision of process or law which can take years to implement. In general, the law is not self-correcting in the ways that science tends to be.

Therefore, unlike the axis between the polarized ideologies of socialism and libertarianism, here I am attempting to explore an orthogonal axis which intersects the competition between those old, traditional ideologies. Running right through them at right angles is realism. This says, for example, that the prediction of what the weather will be tomorrow is insensitive to whatever ideological arguments might be made about the weather in a political context. Only the most idiotic of politicians would attempt to change the weather – although that is exactly what Donald Trump did attempt when he tried to influence hurricane forecasts – but there are plenty of more subtle examples where people apply ideology to Nature in the expectation that Nature will conform. There were many instances of this during the COVID-19 pandemic. When, by chance, these beliefs do align with Nature, this is then seen as a miracle or it is used selectively to suggest a true causal relationship between an ideology and a natural process – the equivalent of the hand of God. It then gets added to the hocus-pocus of belief. But as we know, correlation does not mean causation and selective correlation is hocus-pocus. Selective correlation conveniently ignores all the evidence where correlation does not occur, but for people following strong ideologies, all they search for is reinforcement of their hocus-pocus, like looking for all the positive evidence to support a proposition rather than acknowledging the extent of ignorance. Realism then becomes reinvented in the image of their own ideology. Everything ends up on the axis of ideology, and this, as we know from experience, is a very dangerous place to be. When reality bites, as it inevitably does eventually, it comes at an immense price for people subjugated because of the hocus-pocus.

The virtual graph I have drawn in the last few paragraphs is intended to create a rational expression of where specific issues might lie in the space between idealism and realism. The axis of scientific realism is really an amalgam of two different flavours of realism: the empirical realism associated with Francis Bacon and the rationalism of René Descartes. Bacon's empiricism gives us verified data with which to assure ourselves of reality, and Descartes' rationalism gives us theory about how unobserved, or unobservable, parts of the world actually function, which is inferred from patterns and generalities found in the world as we see it. One involves deductive approaches to reasoning whereas the other involves inductive approaches. Both are methods of exploring the vast expanse of the unknown and of creating a picture of reality. However, if they are not applied with sufficient tough-minded rigour, even empiricism and rationalism can tend towards idealism. A common example occurs when rationalists create models of the world, often rooted in mathematics, which they fail to test adequately against reality, and which they then tout as reality. Sometime theoreticians, as they are called, become convinced that the world they have created within computer models, which are based on logic and are undoubtedly beautiful to them, *is* the real world. This is just as dangerous as allowing every issue to drop onto the axis of ideology. Like any ideology, models need to be tested against reality and shown to be a reasonable fit.

What I have suggested is a Cartesian diagram showing how the different forces acting within the politics factory are tensioned against each other. Different people play their parts in creating the tensions, but it is easy to see if there are no people in the politics factory tensioning the Cartesian plane in the direction of realism, or if these people are corrupted by what is going on inside the politics factory, everything falls towards the axis of ideology. The diagram, therefore, is a practical way of imagining the inclusion of the expanding and increasingly important field of scientific discovery within politics. It shows how, at least in theory, it is possible to exist in the politics factory as scientists and at the same time to retain a clear definition of one's function.

My depiction of the tensioning between ideology and realism was also described by Plato in his allegory of the cave.[6] The allegory explains that we are like prisoners held in a cave and chained for our whole lives so that we face the back of the cave. Behind us in the tunnel leading from the entrance to the cave is a fire and our captors walk up and down between us and the fire, casting their shadows on the back wall. Here, the back wall is the equivalent of my axis of ideology and the tunnel is the equivalent of my axis of realism: one reflects only an abstraction of reality whereas the other, if pursued to its end, allows reality to be directly observed. According to Plato, the moving shadows are all that most people see of the world. For them, the abstraction created by the shadows becomes reality because it is all that they ever experience. People chained within the cave may imagine what is causing the shadows, but until they actually experience the real cause, they will be none the wiser. It is structured enquiry, science, which provides us with the capacity to move beyond just watching shadows. Scientific enquiry unlocks us from our chains. As William Blake put it, 'What is now proved was once only imagined.'[7]

Many people are happy to live their whole lives chained and watching and believing in shadows, but some start to ask whether the reality we imagine is correct. That we see the world through the eyes of the idealists first and the realists second is something which philosophers have grappled with throughout history. We may be born as idealists with some of the instinctive capacities to cope with the uncertainties of the world, but we need to learn in order to become realists. We vary greatly in the extent to which we make the journey along the passage of the cave to displace instinctive or culturally inherited beliefs with knowledge. As the existentialists would say, we exist before we become who we are. Consequently, realism is continually under assault and often seems to be losing out. In the words of Kissinger, realism is very much a minority view.

The result is a constant slide towards subjectivity. The quest for objectivity is equivalent to trying to maximize realism, but objectivity is like a ball on a steep slope and it takes continual attention to stop it rolling back to the base case of subjectivity. When the whole process becomes institutionalized, those participating hardly realize what is happening. Subjectification becomes the norm. James Feibleman called this the 'subjective by-pass',[8] because it is the easy way to avoid the hard work associated with sustaining objectivity. Scientists have to wear the sociological equivalent of a hair shirt to continually remind themselves of the chasm of subjectivity which lies on all sides. It is so easy to fall off down into the abyss, and scientists themselves often do. As a Chief Scientific Adviser in government, it seemed like I was permanently having to push large loads of subjective baloney back up a hill towards a more objective end point, the equivalent of pushing issues back up the axis of realism or out towards the mouth of Plato's cave. Without this continual effort, often involving the relentless and irksome application of scepticism, then subjectification happens by default and we all end up back sitting on the axis of ideology, or in chains facing the back of the cave, where even what is called 'science' is a construct of systematic subjectification.

It is for these reasons that individuals who are acting in the interests of objectivity and who champion the progression of argument up the axis of reality would be uncomfortable with finding a home along the axis of ideology. Within this construct, it is hard to understand how scientists can be associated with specific ideological politics such as a political party. For example, the electoral system in the US tends to force people into one of two camps – Democrat or Republican[9]– but it would make little sense for scientists to associate more with one than the other. Even if the people in one of these camps may be more receptive to using scientific knowledge, it would be hard to justify scientists themselves joining that camp.

As a result, scientists are often left outside the politics factory, sometimes frustrated by the lack of mechanisms to make their voice heard. To deal with this, various gateways into the politics factory have been created for scientists to engage with those inside. In the next chapter, I want to examine how well these work.

5

Gateways to the politics factory

In the words of Solly Zuckerman, 'Scientific advisers . . . are essential at all levels of government' and 'no modern government could survive without their help'.[1] For all that science may be essential, however, does the system of delivering science advice work effectively? Are there effective ways of introducing both scientific information and the reductive and inductive logic applied by scientists into the politics factory?

To those located outside the politics factory, looking inside can present a frustrating picture. Geoff Mulgan suggested,

Governments are now even less capable of using high-quality advice, assuming they obtain it. Ministries – such as those for agriculture and education – often have plenty of experts siloed within their own specialties. But teams around executives struggle to weave advice and evidence together. They tend to be small and consumed by 'firefighting'. Politicians are too busy and distracted to do the job of synthesis, and civil servants are usually more comfortable with law and economics than with science or statistics, or the practicalities of implementation.[2]

Even if some parts of the politics factory make an effort to include objective analysis, often this is superficial. For example, one of the *Seven Principles of Public Life*[3] is *objectivity*. These principles are notionally about combatting the corruption which comes from the pursuit of pure self-interest, but they are mostly treated like a signpost and studiously ignored. Without effective enforcement, such signposts are no more than redundant architecture.

Other non-governmental components within the factory make no effort at being truthful at all except when their self-interest is at stake. For example, there is no point in an oil company deluding itself about the location and size of oil reserves, but these organizations often proactively mislead themselves and others about issues which are not in their interests, such as the existence of human-induced climate change. For many pressure groups, including most that are concerned about the environment and social welfare, the selectivity of evidence to support their own cause is an implicit part of the *modus operandi*. These organizations can be so shameless about how they misuse scientific knowledge to promote their agenda that it is even the expected and accepted mode of behaviour which is often never even questioned by the press.

The honest pursuit of truth is also challenged by some deeper problems. As Robert Pirsig observed, the number of rational hypotheses that can explain any given phenomenon is infinite, but the number which is close to true is very small.[4] This means that for those who play at politics, there is an incentive to ignore science because it narrows the range of possible rational arguments which can be applied in any specific situation. It constrains their scope for generating hocus-pocus to suit themselves.

The politics factory is also full of competing voices: lobby groups, interest groups, self-appointed experts, anybody with a grievance, journalists, some people with expertise plus all the professional politicos who focus on their differences rather than their similarities, resulting in the exaggeration and amplification of triviality. In other circumstances, they could be perfectly civil neighbours, but the echo chamber of politics brings out the worst in them all.

Issues erupt out of the ferment of the politics factory in what sometimes looks like random ways. One I experienced was set off by the influx of a disease of ash trees to Britain, known as 'ash dieback', in late 2012. This was just one of many introductions of tree diseases which happen fairly regularly, but for unknown reasons it sparked a wave of fear in the imagination of the public and turned this event into a national crisis. A national emergency was declared and politicians started to make ridiculous statements about eradicating the disease. Having the nation's trees infected by a disease coming from continental Europe probably reflected deep-seated xenophobia. It sparked a process which is sometimes referred to as *social amplification*, which is the disproportionate exaggeration of relatively trivial issues probably reflecting deeper and unrelated pathologies.

Although this is exactly what one would predict about the behaviour of a complex system, it is one thing to theorize about such behaviour but quite another to live through it and, perhaps, even be an influential node in the process of amplification. Ash dieback was first detected in Britain during the first few months of my tenure as a Chief Scientific Adviser when I was still relatively naïve about how much influence I had. An off-hand comment I made saying that 'native ash trees in Britain will gradually die out and be replaced by new species' and 'Britian's forests will never look the same again' appeared on the front page of the *Daily Telegraph* on 7 November 2012. This may have been part of the trigger process for the amplification, which resulted in media scrums outside government offices, pictures of me walking into COBR[5] meetings appearing on the national news and many weeks of having to quell the desire for what seemed like every person and their dog to milk personal advantage from the situation. Even if nothing which has happened since has suggested my statements were not a correct assessment, it illustrated how very easy it can be to trigger non-linear effects inside the politics factory.

But the politics factory does have self-righting characteristics. As Tam Dalyell, the veteran MP and a former 'Father' of the House of Commons, once explained to me, in the politics factory there are key moments when issues rise to prominence, and once they pass, then they are gone for ever.[6] The noise in the echo chamber rises to a crescendo and then falls away as most people become disinterested or, more likely, one crisis is replaced by an even bigger crisis which then attracts attention. Time alone is the healer as the attack pack of the press turns to look for other red meat to consume. This effect of time is probably driven by the innate latency within the machinery of the politics factory caused by boredom in the mind of the public and those playing political games.

It is against this dynamic backdrop that science needs to be able to inject a sense of mature reflection and a feeling of constancy and stability. In more lucid moments, there is an understanding among some of those who are inside the factory that they need to look outside to take their bearings on the real world. This often happens informally, but there are also formal mechanisms, at least within governments, businesses and institutions. For example, *advisory committees* or even official *offices* often exist to do this translation, but technical experts skilled in everything from statistics and economics to nuclear chemistry might also be employed to ensure reality is not abandoned completely. Pivotal among these experts are people called Chief Scientific Advisers.

(Sometimes they go by the title Chief Scientist when they are a part of the executive structure.)

I see these mechanisms and people as gateways between the world of the politics factory and the world of reality, especially the realities of Nature. My role covered areas of food, agriculture, fisheries and environmental policy and included everything from the chemistry of plastics to the dynamics of fish stocks and from the top of the atmosphere to the deepest parts of the ocean. But sometimes it felt a bit like being an acute trauma physician at an asylum for people who self-harm.

Scientific advisory systems exist to inject realism and expose flummery in the politics factory. They rely heavily on the integrity of processes used to summarize scientific information. One important example is the processes used to collect official statistics, which are designed to provide information about the true state of things.[7] These statistics are generated using rigorous methods and governed by layers of oversight which are specifically designed to stop them being manipulated for political ends.

However, anything which has importance is subject to subversion once it enters the politics factory. Some pressure groups and industry representatives inject their own biased research results into debates. Their tactic is to cloak themselves in objectivity in an attempt to lure people to accept their perspective. Bombarding people with biased statistics is a common subversive tactic[8] because it creates confusion about genuine, reliable scientific advice and stokes the tendency to mistakenly find meaningful connections between unrelated things.

Trust built on a foundation of openness and truthfulness is the essential ingredient of any scientific advisory process. Those people who operate collaboratively within the politics factory in the mode of 'brokers' are likely to be most effective, but they have to also find ways of stopping themselves becoming corrupted by association. People with specific roles like Chief Scientific Advisers need to have the mechanisms to walk this narrow line between maintaining trust and becoming absorbed into the ways of the politics factory.

Walking this line is difficult and fraught with hazards. During the COVID-19 pandemic, in the UK, we saw the vision of Chief Scientific Advisers standing shoulder-to-shoulder with senior politicians who recited the mantra 'we follow the science' when, in fact, those science advisers standing there knew this was not the case.[9] It might have been more accurate to say that policy decisions were being informed by science, but it was part of the disingenuity of the political playbook to enforce science to align with political priorities in order to gain the trust of the public by placing trusted scientists alongside politicians. Should these advisers have appeared in this way? Perhaps it was the only way to genuinely get their advice heard by the public, who, in this case, were the people who really mattered.

Nevertheless, these advisers can be critical entry points for science to influence within the politics factory. Knowing how to influence requires political intelligence and knowledge of the factory's machinery. For example, government has special gateways to the influential centre of politics which tend to be carefully guarded by civil servants, who act as the gatekeepers. Knowing the right people to contact and when to use the gateways can make a difference and reduces wasted effort hammering on the wrong door at the wrong time. Chief Scientific Advisers are present to help guide the way in which science can enter. Peter Gluckman and colleagues have referred to them as 'boundary actors' and 'knowledge brokers'. According to them, 'The interface between science and policy is complex, but can be bridged by ... knowledge brokers, who translate the different languages of the two communities and align information needs and outputs.'[10]

Whatever the descriptor, the job of these people is to help recognize when all the stars are aligned, meaning that the gateway is both open and the right route to take. The political scientist John Kingdon suggested these alignments call attention to issues which open political opportunities, and those who are best able to read those events and maximize the opportunities from them will be most successful.[11]

Apart from knowing the gatekeepers, Chief Scientific Advisers need to have certain attributes. These include the capacity to explain complex issues concisely and to frame technical problems accurately but in ways which are relevant to the context. Self-effacing honesty is also needed, especially about scientific uncertainty, and this can create challenges to credibility. Some people in the politics factory (who may not have experienced life outside its walls) may never have had a serious conversation in their whole lives with somebody whose purpose is to champion objectivity rather than to look after some sort of self-interest. The Chief Scientific Adviser needs a capacity to empathize, although not necessarily sympathize, with the problems and capabilities of the recipients of their advice, especially when this involves government ministers and officials.

Some recipients of advice could not cope with having to deal with somebody who was not being political. I found that some politicians were incapable of thinking in the abstract or theoretically. I shall never forget the horror on the face of a senior cabinet minister when I tried explaining something in theoretical terms. The minister was incapable of thinking in the abstract and was utterly lost and out of her depth. At the opposite extreme, on another occasion, I found myself recounting the paradox of Schrödinger's cat to a minister who was genuinely interested in the meaning of reality and uncertainty. I have also worked with ministers

whose political antennae were sharply tuned to identify scientists who were really pushing some sort of ideological
agenda. They were right to do so, because some scientists
were incapable of distinguishing ideology from objective
analysis.

To be effective, scientific advice needs to be carefully
packaged in ways which reflect *certainty, impact* and *practicality*, and there is a need to exercise judgement based on
these three criteria. Certainty concerns the strength of the
scientific knowledge or the firmness of the knowledge upon
which the advice stands; impact is about the quantum of
effect the advice is likely to have (issues which could have
high political tension but have small impacts are hardly
worth bothering with); and practicality is about the political
cost as well as technical practicality. It is pointless wasting political capital on small-scale issues or issues which
are unlikely to be technically feasible. Getting the balance
among these right requires inside knowledge of the politics factory and skill, without which science will have little
impact.

This is a rosy picture of the Chief Scientific Advisers
working the system to ensure that science has impact, but
nothing is so simple. It was supposedly Churchill who said
that 'scientists should be on tap, not on top', which is just a
reiteration of the idea that the politics factory can keep on
manufacturing irrespective of what science might have to
say. For many working on the factory floor, science will be
a threat rather than a solution, so the presence of a Chief
Scientific Adviser is not always welcomed.

One of the most senior civil servants in the UK observed
to me that it was one job of the civil service to keep science
advisers in their 'boxes' and only to open the lid when they
were needed. At their best, science advisers are completely
integrated into the decision structures in government; at
their worst, they are simply excluded: the 'box' is never

opened but is exhibited as bogus evidence that the government takes science seriously. I have seen examples of both, but mostly it is a mixture of these two states. Those who control the lid of the 'box' then control whether the voice of science is heard within the political echo chamber, and this is a position of power held by civil service gatekeepers. One departmental Chief Scientific Adviser told a House of Lords Select Committee in 2011, 'Sometimes you are denied when you think you ought to have access, . . . but I think part of the job of a CSA [Chief Scientific Adviser] is to make sure they kick the door down, frankly.'[12] While this kind of muscular approach to breaking out of the box does happen (although I do not think doors are literally kicked down), if the relationship between a science adviser and colleagues in government has descended to this level, then trust has disappeared, redemption is impossible and very little can be achieved.

But the 'boxes' certainly do exist and can take many forms, some of which are so subtle that even the scientists who are being contained are not really aware of what is going on. I certainly recognized that the loyal civil servants who ran my office were there not only to support me but also to contain me. Containment involving building the walls of the 'box' is a diffuse invidious process which happens without any overt signalling. Advisers can just not be invited to the most critical meetings, and when this happens, they can be unaware of their exclusion. They can be confined within a set of rules about how they operate, and within forums created for science itself to echo within, like the Intergovernmental Panel on Climate Change (IPCC)[13] and many other technical or expert advisory committees established to deliver scientific advice. They can be given diversionary work to humour them and to soak up their latent energies. When I was a Chief Scientific Adviser, I had responsibility for almost thirty different science advisory committees of various sorts involving around three hundred scientists at any time. One (the Advisory Committee on Fish Names) had never met and, as far as I could discern, had no members. Again, it could have been straight out of a script of *Monty Python*, but it was there for a purpose, although that purpose was nothing to do with listening to scientists. Only a very small number of these committees actually did anything useful or had any influence, but it was in the government's interests to keep them going because the government needed to be seen to be taking science seriously and giving scientists something to do was important. I spent seven years trying to reform these committees but failed because they were useful — not normally for the advice they provided but because of the political cover they gave. Just occasionally, however (such as during national emergencies), a few of these committees can become a godsend to the politicians. The irony is that if the politicians had made better use of these committees, then things like national emergencies would probably be much rarer events.

Access to the politics factory tends to be highly valued by scientists themselves. This means that rationing of access through the gateway is a powerful tool for the politicians and civil servants to create alignment of scientists with political needs. My failure to reform the advisory committees was because none of those who were on the receiving end of advice wanted to cede control of the gateway. If they were going to get advice, it would be on their terms, and, ideally, it would say what they wanted it to say.

Bob May, the UK's Chief Scientific Adviser between 1995 and 2000, thought the Churchillian view of science's relationship with politics was ridiculous. In his view, the fault for not being attentive towards the messages from science mainly lay with those within the politics factory and that they had a duty to listen. He wanted to break the control exercised by the gatekeepers by shifting the rules. He said the political classes needed to listen to the voice of science and be accountable when they chose not to.[14] But his efforts to embed these principles have been almost universally ignored. What he had probably not fully appreciated was the level of counter-measures which had developed across the UK civil service to contain any rise in the power of scientists. The experience of Anne Glover, who became the first and (to date) only Chief Scientific Adviser at the European Commission, was that she encountered the insurmountable wall of resistance to any rise in scientific power within that bureaucracy. As a result, the post was abolished after only three years.

The story is largely the same for the Presidential Science Adviser in the US. This role developed slowly following James Killian's appointment in 1957, reflecting the ambivalence rather than outright enthusiasm about the role of science close to the centre of decision-making. The role was only supported by a small staff in the White House Office of Science and Technology Policy,[15] a design intended to clip the wings of science advice – and to minimize the size of the gateway to the echo chamber. Giving scientists a little symbolic recognition was appropriate in order to prevent them accruing more power.

Peter Gluckman, another experienced scientist operating as a senior adviser in government, this time in New Zealand, attempted to construct a more subtle set of strategies for interaction between the scientific and political systems,[16] referring to the process as an 'art'. It is hard to disagree, but should such a critical function be left to something so intangible as an art form?

In this battle of tit for tat, the scientific community has responded. In the quotation at the start of this chapter, Geoff Mulgan referred to 'synthesis' as an example of a tool used to open the portal to the politics factory. Perhaps partly in response to its exclusion from the politics factory, the scientific community is becoming much more organized when it comes to helping to deliver important messages to people who find it hard to listen. The reports of the IPCC are an example of *synthesis* and a relatively new tool for influencing is also the production of *systematic reviews* of science in key fields which are important in a policy or political context. Recognizing that it is easy to produce scientific syntheses, or reviews, which have strong biases for preferred messages, these systematic reviews use defined methodologies to synthesize among many different scientific studies with the aim of simplifying the sometimes bewildering array of potential messages, and doing this in ways which are unaffected by the identity and beliefs of those carrying out the reviews.[17] They are part of the quest for objectivity and were used to boil down a very fuzzy picture of the emerging pattern of disease during the COVID-19 pandemic.

Reports produced by august institutions like the Royal Society or the US National Academy of Science can also have a similar capacity to synthesize and lead government into thinking beyond the boundaries of the political echo chamber. They can bring reality to the politics factory about issues as diverse as human–machine interfaces, carbon capture and storage or the future of land use. It is often unclear how much influence they have, although just occasionally they hit a sweet spot. One report I co-produced with Mark Walport, titled *From Waste to Resource Productivity*, formed the basis for England's waste strategy.[18]

One important characteristic of science advice, which is often missing from sociological analyses of how it works, is that of trust between the adviser and the recipient. As C.P. Snow observed with respect to high politics, 'The most obvious fact which hits you in the eye is that personalities and personal relations carry a weight of responsibility which is out of proportion.'[19] This is why having a person like a Chief Scientific Adviser working the politics factory floor and building personal relationships is essential if science is to have any influence within the current factory machinery. Otherwise, by default, science will be excluded because the machinery has never evolved to be inclusive of either science or reality.

High politics is an intensely personal affair. It forces subjectivity on every aspect of decision-making because it is often less about the issue than about the person dealing with the issue. This is why, in my first meeting with any new minister, my opening comment was, 'I am here to help you.' It is also why a minister once looked seriously at me after I had given a supposedly neutral presentation on a technically difficult subject and asked, 'But what do *you* think we should do?' He was less interested in the facts in the case, which were never going to resolve the dilemma at hand, than what I, as an ethically sensitive and relatively well-informed human being, actually felt was right. He was interested in me, the person, not the analysis I had brought to the discussion. That is a powerful position to hold, which one needs to recognize and moderate very carefully. Telling truth to power is one thing, but telling power what it should do is entirely another.

Therefore, most of all, the science adviser needs to gain the trust of those on the receiving end of advice, whether this is ministers, civil servants or the public. Trust is built by listening, communicating thoughtfully and being mindful of only intervening when it is possible to add value. Above all, there is a need to understand when to stand one's ground and when to give way; when to tell the minister that what they are saying is 'rubbish' and when not to. There were many issues I saw which I thought had crossed an ethical boundary because they were not supported by appropriate scientific knowledge or interpretation, but I had no wish to expend my own political capital on them. That was reserved for important issues at key moments when my advice might have impact. Those moments can be tense, and only once did I bet all my chips on an issue which I thought was worth the risk. The issue involved potentially publicly disagreeing with a ministerial decision because it clearly contradicted the scientific advice, and in this case I felt the public was entitled to know what the scientific advice had been. The fact that I survived shows that it paid off, but it also shows there is a need to be willing to gamble.

Science advisers need to be respected scientists with a considerable depth of intellect, experience and a prodigious breadth of knowledge about science. One is useless if when asked for an opinion one demurs because the issue is outside one's expertise. I dug deep into my basic training as a scientist, and it is remarkable how much those basic skills can play to a very broad range of subjects. One needs to read the scientific literature – in my case, the weekly journals *Nature* and *Science* (and I stayed on as a Reviewing Editor at the latter partly in order to stay abreast of the latest scientific knowledge and opinion).

One also develops a nose for false information, the spinning of dreams by scientists themselves, and pseudo-knowledge or claims which are touted as 'scientific' but are either over-blown or possibly downright fraudulent. This reflects a capacity for deep but rational scepticism, critical self-awareness and an appreciation of the basic principles of how evidence is constructed, of what the philosopher Charles Sanders Peirce called the *pragmatic maxim*[20] and which his friend William James described as *tough-mindedness*.[21] There is a need for a T-shaped intellect – at once narrow and detailed, diving deep into a specialist subject but also with shoulders broad enough to engage at more than a superficial level across a very wide range of subject matter. For example, Vannevar Bush is said to have had the uncanny ability to flit from one subject to another and to sound credible and comfortable with all of them.[22] Perhaps the science adviser also needs to be a policy entrepreneur by matching scientific progress with opportunities for policy evolution and for ensuring politicians are aware of these opportunities.

Scientific advisers come into their own when the realities of nature come to visit a political and societal system which is otherwise disembodied from those realities. This is what happened when COVID-19 struck in 2020, but similar, even if less spectacular, examples (such as flooding, incipient dam collapse, deployment of a chemical weapon, contamination of food and introduction of a new deadly tree disease) happened roughly annually during my watch. People in the politics factory suddenly flipped from indifference about science to having not nearly enough of it.

The relationship between science advisers and senior politicians can make a difference, but this comes with risks. Commenting on the influence of Frederick Lindemann (Churchill's science adviser during the Second World War),[23] C.P. Snow was cutting: 'It is dangerous to have a solitary scientific overlord.'[24] To counter this, I had a Science Advisory Council of seven eminent scientists to look in on what I was doing. If science advisers go native, their usefulness diminishes and the subversion of science is accelerated – this is what Snow saw in Lindeman. In my view, science advisers should not have executive responsibilities within government because this compromises their role as people who can challenge the norms within government. Ideally, they should also not depend on being judged purely from within government for their performance and career advancement.

The 'art' to which Gluckman refers includes the capacity to retain the trust of those around you while standing astride the gateway with one foot inside the politics factory and the other outside to stop the descent into corruption.

This all sounds a lot like Weber's description of politics as an occupation involving a pragmatic mixture of aspiration and practicality, making it the art of the possible. So much of the existence of a science adviser in government is about striking a balance between just going with the flow of the politics factory or being challenging about its internal machinations with the effect of being shut out, confined to the 'box' and therefore potentially having to 'kick the door down'. Being present at the table when key decisions are made is always better than sitting outside the room and having to rely on those inside reading and understanding a hefty tome in the form of a scientific synthesis report. The problem with most science advisory systems is that this physical and personal presence is still by invitation, not by right.

If we see science as a central plank supporting the health and welfare of people and which is evolving along with society itself, this seems to be an incredibly fragile way of working. When Patrick Vallance said in 2022 that scientists should 'stay in their tramlines', this was not because he wanted to silence or constrain scientists but because he knew how difficult it was to ensure that the voice of science was heard at the table where key decisions are made. I saw myself how well-intentioned but insensitive and sometimes ideologically motivated interventions by scientists made it harder for that voice to be heard in the right places and at the right times.

6

Shoring up the marriage

Many countries have adopted a system involving single individuals, Chief Scientific Advisers, who can work at the centre of government to translate the language and culture of science into that of politics. These include Australia, the Czech Republic, Estonia, India, Ireland, Latvia, Malaysia, New Zealand, the UK and the US. Others, like China, France, Germany, Italy, Japan, South Africa, South Korea, Switzerland and Taiwan, have slightly different approaches, often involving links between leading science institutions and the national or regional governments. All this could be seen as an effort by the political elite to embrace science, but it is a fragile response.

In Canada, the post of Chief Scientific Adviser was abolished by the conservative Prime Minister Stephen Harper in 2008. Science was literally turfed out of the politics factory. It was welcomed back in by the liberal Justin Trudeau in 2017, but this exemplifies the knife-edge vulnerability of the voice of science within politics as well as the implicit politicization of science, emphasizing that one ideological group is more accepting of science than another. I once had to tell my Science Advisory Council to tread carefully because the new Secretary of State was of a mind simply to abolish them if she felt they were going to be a bundle of trouble – a thinly veiled process of oppression and coercion.

There would, therefore, appear to be a *prima facie* case for mandatory inclusion of science in the machinery of the politics factory as a constitutional right, an option considered recently (although ultimately rejected) by Chile. Without more robust representation of science within the politics factory, then all we have is a marriage which will surely hit the rocks again sooner or later.

Many international organizations, such as the World Bank and the Food and Agriculture Organization, also have Chief Scientists. Chief Scientific Advisers within departments in the UK emerged from a review of science carried out in 1971 by Lord Victor Rothschild. The then minister responsible for science and education, one Margaret Thatcher, decided to privatize government science, leaving government in need of a person who could act as an 'intelligent customer' (a tacit admission that government bureaucracies are incapable of using science without help).[1] In South Africa, the role of Chief Scientific Adviser can be regarded as the Minister of Science and Technology, who is then supported by specialists in different fields. The UK has also experimented with making a former Chief Scientific Adviser into a minister. But these ministerial roles potentially confuse advice about *policy for science* with a role in providing advice about *science for policy*. Chief Scientific Advisers certainly have an important role in promoting the strategic positioning of science but they function mainly to build better government and better policy rather than as advocates promoting interests of science and scientists.

Chief Scientific Advisers can, on occasions, change the course of history. On 6 May 1940, while Europe was falling to the Blitzkrieg, Vannevar Bush had a fifteen-minute meeting with President Roosevelt in the Oval Office when he presented a single-page proposal to establish what was to become the most effective scientific production system that has ever existed. This was the academic-industrial-military complex which operates in the US and which can probably be credited with powering, for both good and bad, the late 20th-century industrial might of that nation. The US government spends prodigious amounts of money on scientific research, especially through the portals of defence, agriculture and energy, as a direct way of stimulating economic growth and development. Bush realized that this involved creating a partnership between government and industry where government picked up the costs of research which was too early in its development to be capable of making money and it justified this politically because of the need for military and civil security. *Policy for science* was born because science became a critical strategic asset.

Bush was, in the process, partly responsible for developing what could be called the *British/American Model* for a science advisory system. This focusses on placing key scientists close to the seat of political power. It contrasts with what could be called the *Continental Model* of advice (because it is common in Europe). Advice, in that case, is most often delivered using a committee-based structure where scientists are asked questions from within the politics factory and they provide an answer often by consensus in the form of a committee report. In a similar way, the Organization for Economic Cooperation and Development (OECD) has also set out a technical process for providing scientific advice for policy,[2] as have many of the international organizations established to manage trans-boundary environmental agreements. These often rely on advice from scientific committees which report to the organization's politically based governing body. The United Nations has also made several attempts to establish a science advisory system similar to that of the OECD, and the European

Commission has flirted with creating a system. It first attempted something like the *British/American Model* – by appointing a Chief Scientific Adviser – which failed in the wake of infighting among the power structures within the EU. Eventually it seemed much more content with the *Continental Model*,[3] mainly, I suspect, because it could be controlled and scientists could be kept in their 'box'.

Both these systems have strengths and weaknesses. The *Continental Model* has a rigidity of procedure which might provide assurance of quality, but anybody who thinks this should consider the disastrous history of the Scientific Committee of the International Whaling Commission as a worked example of just how badly this can go wrong.[4] The scientists nominated to the committee were those who were already indoctrinated in specific national interests, and the result was the near-total extermination of most of the global populations of large whales. The same happened in most fisheries management organizations around the world throughout the 20th century, resulting in chronic over-exploitation.

The *Continental Model* also lacks the personal touch, involving the translation of the dense, dreary, sobering objective reasoning often contained within dry reports like those produced by the IPCC into subjective reasoning involving the authoritative personality who carries the messages in ways with which the political classes can engage. The question being addressed at the point of decision-making is: 'How much political risk am I taking if I make a particular choice?' Having a scientist in the room (or, even better, having a scientist sitting in the role of decision-maker) at the time when decisions are being made can influence the flow of conversation and the construction of reasoning. Just by being present in the rooms where decisions were being made, even if I said nothing, was enough to shift behaviours in others present. The *British/American Model* has much greater fluidity and flexibility but carries the risk of greater reliance on the attributes of the individuals occupying the advisory roles (see my description in chapter 5 of the danger this can create using the example of the Lindemann case).

Both systems have evolved in recent decades to value the 'independence' of advisers from political influences. The issue and definition of 'independence' is complex and controversial because it is a target of powerful vested interests. For example, the tobacco industry, beginning in the 1950s, intentionally created industry–academic conflicts of interest to muddy the waters around the independence of scientific advice and its validity.[5]

'Climategate', which emerged in 2009, was a similar action by vested interests to undermine scientific advice about climate. Illegally hacked emails between scientists were selectively spun in a way to deceive and discredit climate science.[6] Pressure groups, or special interest groups, also known as NGOs, often selectively promote experts who support their own agendas but, in reality, they are business operations whose product is disruption or pursuit of a cause rather than objective analysis. Nothing they promote could conceivably be *independent*, even if that is how it is usually disingenuously presented. Unlike industry, commerce and government, they often enjoy a privileged position by largely escaping journalistic scrutiny. They even sometimes establish outwardly legitimate research groups whose task is to carry out research and publish it in the scientific literature, all the time tilting the scientific narrative in the direction of their paymasters. Creating destructive ambiguity about the 'independence' of scientists is a potent tactic used by all those operating within the politics factory.

The European Food Standard Authority (EFSA), an arm of the European Commission, goes to tremendous lengths to weed out scientists who have the slightest whiff of non-independence in the form of an apparent conflict of interest, but thoroughness of procedure, applied in bone-headed ways, plays into the hands of those who wish to undermine scientific advice. By letting research contracts go to the most informed scientists in a field, agrichemical companies, either by design or as a consequence, disqualify these scientists from advising on what regulations are appropriate. Having been captured by vested interests, the skills of these scientists are no longer available to act in the public interest and the result is that EFSA does not always have access to the best scientific advice.

Procedural thoroughness is no substitute for individual integrity. The reversion to procedure probably undermines the basic ethical conventions of individual accountability and responsibility for one's own actions. It is really impossible to tell whether a scientist has been 'bought'. Even scientists themselves can be unaware of being targeted, such is the subtlety of what goes on when they step into the politics factory.

As the political philosopher John Rawls said, people are rarely, if ever, disembodied or disinterested parties in almost any circumstances,[7] and despite some fundamental differences of views, the contemporary philosopher Peter Singer agrees.[8] Whatever their race, gender, financial interests (even to the extent of apparently having no money) or religion, individuals have interests. Rawls described ways of dealing with this by allowing conflicts to lie behind 'the veil of ignorance' – meaning that in the case of scientists, when making judgements, it is best for them to be ignorant of the political and social circumstances in which those judgements might apply. When advisers are ignorant about the interested parties affected by their advice, then their advice is more likely to favour the greater good[9] – or perhaps it is just likely to end up supporting their personal interests? The philosophical point is that one can strive for independence of advisers from vested interests, but it is rarely achievable.

Scientists who proclaim themselves as *independent* are probably not being honest with themselves about where their conflicts lie.

During the COVID-19 pandemic, some people thought that the UK would be better off getting its scientific advice from an 'independent' group of scientists.[10] They were referring to the fact that the UK government was using a range of scientists from across academia as well as public sector institutions to compile scientific advice, and out of practicality this could not include every scientist with an opinion.[11] Similarly, in the US, there was an appeal for more *independent* scientists to advise on pandemic response.[12] Diversity among those giving scientific advice is certainly important, but it is also common for those who disagree with scientific advice to question the so-called 'independence' of that advice. There is also a common misunderstanding that an 'independent' scientist is any scientist who has no association with government. This conveniently forgets that the politics factory spreads its tendrils to much more than government, and many of those scientists themselves calling for more *independent* scientists to advise on the COVID-19 pandemic already had deep roots in the politics factory and failed even to recognize, let alone declare, those interests.

One emerging generality is that calls for greater independence in scientific advice are usually subversive acts designed to either undermine scientific advice or load the scientific jury in ways which are more likely to match a particular ideological position. There is certainly a need to ensure that individual scientists providing scientific advice are acting ethically and in good faith. There should be an expectation of ethically centred reasoning on the part of scientific specialists, and such expectations need to be recognized and regulated. Systems of accreditation allowing scientists to practise as professionals – which can be both given and withdrawn – are needed much as they exist in professions like the law and medicine.

The OECD also suggests that scientific advisers should carry liability for their advice. In the past, a few attempts have been made to prosecute scientists who have appeared to provide bad advice, although these rarely succeed.[13] The OECD considers that expert advisers operating within an official mandate from government should be accorded a status similar to that of government officials, meaning that it is the government rather than they as individuals who would be held liable for the advice provided. Much of this has not been tested thoroughly in law, and government officials in many continental European countries, where there are systems of civil law, appear to have much greater individual liability for their actions than is the case in places like the UK or the US, where common law applies. International law also currently makes it difficult to launch lawsuits against bodies like the IPCC, but litigation is an issue which few scientists consider. Whatever the legal risk, it appears that not using systems of accreditation leaves us uninformed about the quality of scientists who provide advice. The cynic would say that is exactly how many of the players in the politics factory would wish it to be, so that they can pick and choose the people who are most likely to provide the advice they want.

The civil–common law differences in liability risk may be slight, but they present a subtle shift in the balance of risk for scientific advisers in different legal systems. This shift may explain some of the reasons for the difference between adoption of the *British/American* versus the *Continental* models of science advice. If individuals like scientific advisers within government are personally liable, as it seems they are more likely to be within the *Continental* system, then this will inevitably lead to risk aversion. Relying on systems of official committees rather than individuals placed within government possibly reflects a way of managing legal risk. It may also be the reason why precaution has progressively worked its way into European legislation on issues like food and the environment.[14] While the idea is speculative, risk aversion among scientists, depending on the legal risks they perceive, is likely to play a part in framing scientific advice.

These different cultural perspectives were probably at play when the World Trade Organization responded to a complaint by the US by finding against the European Union in its decision to ban the importation of genetically modified (GM) food products.[15] This stand-off on GM food continues to the present day, even though both the US and Europe have used the same scientific evidence to support their positions. Their scientific advisers have come to different conclusions mainly because the advice from scientists in Europe has been more risk-averse. These differences are also likely to drive divergence between the scientific advice given in the UK and the European Union following Brexit. Whereas the UK has a tradition of using risk-based assessments which balance the level of knowledge and understanding with the amount of risk to be taken – something known as proportionality – Europe has a tradition of taking an approach which sees hazards as something to be prevented and which is skewed towards applying precaution. These differences in perception, though subtle, are built into how the different legal codes are used. They infiltrate scientific discourses and affect advice, thus progressively tipping the balance of decision-making and setting different systems of governance on divergent tracks. Scientific interpretation based on the social climate around the scientists doing the interpretation can, therefore, result in intractable political tension.

In the end, I am driven to ask the question whether any of these science advisory systems really work in the ways they should. Ban Ki-moon, the UN Secretary-General, created a science advisory board in 2014, but a follow-up assessment by his successor, António Guterres, found that it had no influence on policy or decision-making. Basically, it was useless, and I suspect much the same could be said for most science advisory systems if they were honestly assessed, perhaps with the exception of their influence at times of national emergency. Even the reports of the IPCC struggle to be effective. The idea of placing scientists in a 'box' and then ignoring them seems to be a strong *modus operandi* of the UN, but it is especially prevalent in large government bureaucracies. Following the COVID-19 pandemic, a Canadian commission which reviewed how scientific evidence had been used recommended that 'Every national (and sub-national) government should review their existing evidence-support system (and broader evidence infrastructure), fill the gaps both internally and through partnerships, and report publicly on their progress.'[16] Perhaps we have relied for too long on science being a kind of extension service bolted on to extant systems of government. Making science work within governments, as a prelude to changing the cultures within the politics factory, probably needs root-and-branch reform of the machinery of government itself. But, even more, it needs a reassessment of how science is used and abused within the politics factory. To illustrate this, I now turn to the abuse and corruption of science through its contact with politics.

Part 2
Science corrupted

The conventional view of science from within the politics factory is that while science brings many benefits for society, it can be a significant source of injustice because not all science results in benefits and some of its products can be very damaging indeed. While evidentially true, in part 1, I have challenged the notion that this is an innate characteristic of science itself. Rather, it happens because of how science is procured and used, which is determined by politics. It is exacerbated by the systematic exclusion of scientific thought from, or its selective inclusion within, the political process. Some of the behavioural pathologies which result from this can be exhibited by scientists themselves, but if bad things happen as a result of scientific progress, there is rarely a thought given to the idea that perhaps it is the way we live and the way in which we use this knowledge which is at fault and needs to change, rather than the existence of the knowledge itself.

It is also a common trope among people who reside within the politics factory to shift blame elsewhere for the existence of inconvenient truths. One of the most potent ways of managing the contact zone between science and politics in order to avoid these inconveniencies is to construct it so that only the convenient truths emerge. The problem this creates, however, is that the more that the inconvenient truths are allowed to reside in the void we call ignorance, the more they hit back at us in unmanaged ways. They become devastating surprises, such as the pandemics of the future.

In part 2, I want to consider how politics bends and corrupts science to focus in on convenient rather than inconvenient truths, or even to create untruths with the illusion of truth. Scientific research is a social process as much as anything else which is the product of humankind, and it is therefore subject to many corrupting influences, but this does not mean that the pathologies which grow as a result within the scientifico-political system should go unchecked or unrecognized. It takes a lot of hard graft, guts and determination to prevent this corruption from completely undermining science. But my central message is that to overcome this corruption, it is essential to apply what I call rational scepticism, combined with a heavy dose of tough-mindedness about the rigour used to judge what is included as valid knowledge.

7

The subjective by-pass

I am going to refer to the pervasive, diffuse, subversive but constantly present subjectification of knowledge within the politics factory as the *subjective by-pass*. This was a term invented, as far as I know, by the philosopher James Feibleman, who said, 'I am going to call it [the subversion of science] the *subjective by-pass* . . . because it meant going around the consideration of the world with all its richness and variety, skirting it, so to speak, in order to get back sooner to a concentration on the self.'[1] Like a nasty environmental pollutant, it turns up in all sorts of places it should never be. I want to illustrate the ways in which it works and how it ends with the corruption of science itself.

One does not need to look far into apparently sensible literature to find a starting point of this illustration. For example, in his hagiography of humanity, Yuval Noah Harari says the following: 'Science is unable to set its own priorities. It is also incapable of knowing what to do with its discoveries . . . scientific research can only flourish in alliance with some religion or ideology. The ideology justifies the costs of the research. In exchange, this ideology influences the scientific agenda and determines what to do with the discoveries.'[2] Harari is questioning whether science can exist outside the social context of its creation. As an activity carried out by people, it obviously cannot, but people often confuse science, the process of discovery in the form of research, with the products of science, and especially with the body of knowledge that it is responsible for creating.

Science as a human activity, just like politics, is ephemeral, but its discoveries are foundational and potentially timeless. The knowledge we have now is to some extent the aggregate of all our cumulative interests throughout history, so it is something more than an alliance with religion and ideology. It is the formal record of our accumulated experience of Nature. People who seek new knowledge (by testing the questions they ask about how the world works against the experience of how it is observed to work) are scientists and, up to a point, this makes us all scientists. This is also why the term 'scientist' is so slippery, because we try to create divisions between scientists and non-scientists which are arbitrary. However, for some people, science is a purposeful activity, and it is this group we most often associate with the title of 'scientist'.

Harari makes science into a subsidiary or servile activity associated with a religion or ideology – science on tap and not on top. This position needs to be challenged and changed if there are ever to be appropriate levels of objectivity within public discourse and to stop the constant pressure to force political issues along the subjective by-pass onto the axis of ideology. There is no doubt that science has been entrained by ideological influences, as Harari suggests, and continues to be – but this does not make it right or inevitable that such entrainment should happen. Turning Harari's argument around, it is worth asking if ideology, or politics,can prevail without science. When they exist without each other, neither is likely to lead to fairness and justice. As I have already said, there is a lot of evidence that separation leads to tyranny.

Harari's view is probably quite common. But subservience of science to ideology leads to pathologies. For example, 'good enough for government work'[3] is a term sometimes applied to something done to a low standard, like the work of a servant who is fed up with an unappreciative, incompetent, miserly or bullying master. In my time in government, I often encountered research done in this fashion, meaning that it was really technically substandard. In some sort of misguided attempt to create an 'alliance' with science, government often commissions science to do its bidding without much appreciation of what to do with the outputs. Reports were produced by scientists who probably knew those in government who had commissioned their work would never even read them or would not have the skills to judge their quality. I saw the same piece of research commissioned multiple times because those involved had forgotten – or never knew – it had been commissioned before. I also saw piles of research reports unread, unpublished and unused. By the time they had been delivered, nobody could remember the political imperative which stimulated them to be commissioned in the first place. This was not just inefficiency, but also reflected the process of commissioning of research as a means to a political end. It was the act of showing concern to obtain new knowledge, of gathering 'evidence', that mattered more than the knowledge itself. It was a tactical way of buying off criticism, and in this context commissioning scientific research became part of the political playbook. There was never an intent to create new knowledge.

I had several tense stand-offs with senior politicians when I insisted the government needed to publish all the research reports it commissioned, irrespective of whether the results supported current political preferences. They then responded by cutting research budgets because they preferred not to have research done at all rather than risk commissioning research which might come up with the 'wrong' answers. This was Harari's suggested subservience of science to ideology in practice. Saying, therefore, that 'scientific research can only flourish in alliance with some religion or ideology' starts to sound quite comical, even absurd. Often precisely the opposite is true. Subjugating the search for objectivity becomes the objective, and this made me conclude that governments are irrevocably conflicted when it comes to commissioning research.

When science becomes the servant of ideology, bad outcomes are probably inevitable. Eugenics, for example, is 'the science which deals with all influences that improve and develop the inborn qualities of a race', at least as defined by Francis Galton,[4] a 19th-century champion of the field. While phenotypic or genotypic differences among people are self-evident, eugenics starts to define these in terms of their quality. The key is in the words 'improve and develop'. This brings subjective values to bear on what improvements and developments look like. Improvement, in particular, is in the eye of the beholder. Eugenics is a product of science becoming an instrument of ideology. By standing back from values-based arguments and allowing others to hold the ring on those issues, and by refusing to make commitments to expose value judgements, scientists are in danger of being entrained by toxic ideologies.

People like Galton who cannot see the problem with this are not uncommon and are complicit in driving the subjective by-pass. When studying for my Ph.D. at Cambridge in the 1980s, I came across scientists who were supporters of *social Darwinism*, a concept which underpins racism, anti-semitism and social pathologies like unconscious bias in the workplace, but also how preferences play out in international geopolitics. Dressing up ideology as objective science helped to mould the toxic politics of national socialism as practised in Nazi Germany but it pops up in other forms of discrimination in the present day, including within science and especially within medicine.[5] Stopping this happening needs a lot of tough-mindedness, but it also needs a recognition that Harari's ideas[6] about the subservience of science simply sustain this kind of toxicity.

If properly used without a heavy emphasis on values, science provides a pragmatic counterpoint to ideology. It is the dispassion and the grip of the real and possible in Weber's bureaucratic theory of public administration involving rationality, hierarchy, expertise, rules-based decision-making, formalization and specialization.[7] Rather than being rooted in religious or ideological dogma, when properly applied, science can shape ethical standards and debates, helping to define how different choices impinge on rights, fairness, freedoms, obligations and benefits. To avoid the subjective by-pass, science needs to be able to question and shape ideology (as it has done, though far too late, in the case of eugenics), rather than simply support it, as Harari suggests.

Harari is also throwing a spanner in the works of something called the *Haldane Principle*,[8] which states explicitly that science should 'set its own priorities', an acknowledgement both of the need for priorities in science to be outside the immediacy of politics, vested interests and ideology, and that, in general, it is scientists themselves who have the technical know-how to judge quality. Most democratic systems have something equivalent to the Haldane Principle, but its critics point out that leaving the decisions to scientists about what matters to society gives them special powers and a lack of accountability. As Harari suggests, scientists cannot isolate themselves from religion and ideology, but the Haldane Principle has the potential to do this. It may be an extreme response to prevent the subjective by-pass, but it is an incomplete answer because it does not stop scientists losing or ceding control of the products of their creativity, thus breaking the connection between the ethical rationale for scientific research and the ultimate consequences of that research. It also sometimes results in scientists prioritizing research which is more in their own interests than those of the public, which is just another pathway along the subjective by-pass.

Fixing the subjective by-pass is, therefore, easier said than done, but there is no doubt that it needs attention, because, as I shall show in subsequent chapters, it has a corrupting effect on science.

8

Products of the politics factory – evidence: *quod erat demonstrandum*

> I've noted that we have moved to putting out 'calls for evidence' on a variety of topics . . . we need to be sure of our ground when it comes to what constitutes evidence. There is a danger that this just creates an incoming avalanche of unverified drivel.

I wrote this in a note to ministers, senior civil servants and ministerial advisers. Government had imagined that it could engage people on difficult issues of policy, and potentially defuse controversy, by asking them for their evidence. The problem this created was that 'evidence' had become all things to all people, everything from rigorous science to hocus-pocus. Anything anybody thought was evidence would be counted as such in the scenario. The subversion of evidence is one of the striking examples of the subjective by-pass at work. Far from being one of the products of science, 'evidence' has become a product of the politics factory, and even scientists have become captured by the evidence bandwagon.

Evidence has found its own lexicon of extensions, such as *evidence-based policy, evidence-informed policy* or *evidence synthesis.* There are even *evidence budgets, evidence special-ists* or *evidence directors* in government, which points to a process by which money is often spent on actively generat-ing ambiguity and people are assigned the job of overseeing this. Of course, sometimes this constitutes a genuine intent to improve knowledge, but when stripped back to its fun-damental rationale, it is most often found to be an effort to support an ideological position. Some social theorists claim evidence-based policy (or even 'evidence' itself) is a social construct,[1] and based on what I experienced in government, it is hard to disagree. There is a danger that 'evidence' has become something with esoteric qualities in the sense proposed by Leo Strauss – that one cannot read it literally because it is a construct which, like the works of classical philosophers, is designed to avoid political clashes. I encountered people who even thought it was justifiable to spend money from *evidence budgets* on cocktail parties for visiting dignitaries from overseas, which I suspect is not what Strauss had in mind, but it illustrates my point. In the language of behavioural science, the generation of evidence becomes a displacement activity – valueless except in the sense that it reduces the probability of conflict.

People inside the politics factory are confused about 'evi-dence' because they fail to distinguish between its legalistic and scientific meaning. In legal terms, or at least in normal usage, evidence is the body of apparent facts suggesting whether a proposition is true or false. The veracity of evi-dence is also a matter of judgement: evidence is reliable if people believe it after it has been subjected to adver-sarial scrutiny. On the other hand, scientific evidence is knowledge used within a particular context: it is something objective applied to a subject. It is generated by a process of hypothesis falsification, or alternatively through consen-sus building about how the world works, based on testing ideas against empirical reality. Therefore, legalistic and scientific constructs of evidence are very different, but they are used completely interchangeably within the politics factory. 'Evidence' starts to mean different things to different people, but the dominant voices in the politics factory mostly promote the legalistic definition because this reflects their cultural leanings. Science is then left to undertake a task for which it is not suited, resulting in most people misinterpreting scientific evidence when it is produced.

If science-based evidence was being generated properly, it would emerge from something known as the *knowledge hierarchy*, a concept which first appeared in the early 1970s[2] as a way of bringing structural integrity to how evidence is generated. Some versions of this are illustrated as a pyramid and present *wisdom* at its peak, rather than evidence, but the difference is probably contextual and semantic.

At the base of this pyramid are ubiquitous *data*.[3] Individual data points then aggregate into *information*, which in turn forms *knowledge*, and when knowledge is contextualized, such as when used to create policy or within political argument, it becomes *evidence*. So that we are in no doubt, let's look at each of these in turn.

Data are the basic measurements from the environment around us – everything from the 0s and 1s forming the bits of data in electronic sensors or the firing of a neuron to the tick in a box of a survey form or a pixel in a digital photograph. *Information* comes from data, like the amalgamation of pixels to form a picture. *Knowledge* is created by synthesis among many disparate sources of information – equivalent to interpreting what is being depicted in a digital photograph.

Data are measurements made using accepted reference standards. This means that data cannot just be anything an observer wants them to be, but they need to be rooted in reality, a measurement of some sort. Data also become fetishized within the politics factory mostly when they are being referred to by people who have never thought deeply about what they mean or who want to use them to further their own interests. *Measurement* – the process of acquiring data – is a large and complex subject and could merit a book in its own right, but the lack of discipline about what constitutes data is probably where most of the problems with *evidence* start. If the data are wrong, then everything else that follows – *information, knowledge* and *evidence* – will also be wrong.

Information is the product of a process of structured reasoning based on verified or calibrated data. Information can often be independently replicated so long as the process used to derive the information from data is known. Often information is derived from data using statistical methods which test the idea that data contain information.

Knowledge is the synthesis of information. Scientific papers published in the peer-reviewed literature tell stories in very formalized formats, and they do this in order to build knowledge from the information they contain. There are also now well-developed methods for constructing knowledge syntheses.[4] Scientists have an especially important role in the creation of knowledge both through formal synthesis and through deliberative processes involving the solicitation of expert advice.[5] Some forms of synthesis known as *systematic reviews*, referred to in chapter 5, are likely to be much more reliable than others because they follow a rigorous methodology.

Therefore, all evidence is ultimately based on data, but the way in which data are translated into evidence can greatly influence what the evidence says. All sorts of errors can creep in and all evidence will contain errors, which means it has a level of uncertainty attached to it. If this uncertainty is not addressed and made explicit, then the evidence should be challenged and probably not used at all. Journalists are especially bad at misrepresenting evidence when they report it because they fail to talk about uncertainty, or to demand this from their sources. Knowledge presented without an expression of uncertainty is like a map being presented without a scale – potentially deceptive and often quite useless.

The sources of errors which create uncertainty can also be broken down and understood. This is a method used by scientists in an attempt to 'parameterize uncertainty' by expressing confidence levels around *knowledge* and the evidence derived from it. Failure to properly recognize *error* leads to mistakes of interpretation which are themselves part of the error-creation process. Error can be parsed out into its components, including *measurement, sampling, process, model* and *implementation* or *interpretation* error.

Measurement error creeps in because measuring instruments or processes are imprecise. Everything from a desktop ruler to an atomic clock, which measures time to an accuracy about 1 second in 100 million years, is subject to error. Tests for COVID-19 were subject to errors known as *false negative* or *false positive* results, and it was essential to include consideration of these in any interventions used to manage the spread of the disease where testing was involved. Measurement error can also be biased towards higher or lower values, resulting in the creation of a false impression – often false negative tests are more frequent in disease control than false positives.

Sampling error is caused by the fact that it is rarely possible to measure everything one is interested in and we have to be content with taking a sample. We are apt to make sampling errors by, for example, extrapolating from narrow experiences and imagining that these apply everywhere. Opinion polls often use a sample of just a few thousand individuals to represent the opinions of millions. They are never a precise prediction of the outcome of elections because they are a small sample of the population they try to reflect. Again, bias can play a part too. Until recently, estimates of the rise in global temperatures were biased towards the Northern Hemisphere because that is where most of the sampling of temperature was happening.

Process error is much harder to understand. This happens because we do not have perfect knowledge of the process we are trying to measure or sample. It is sometimes known as natural variation. For example, some people are more likely to test positive in a disease test than others even when both have the disease. Sometimes process errors are expressed as assumptions stated by those doing research, but often they are hidden because researchers do not even know that they exist.

Model error arises because we are all continually testing what we sense about the world around us against some form of mental model of the world. We are asking questions of the world and testing whether the questions are right or wrong. As psychology has taught us, the way the question is asked affects the kind of answers which come back. The question 'To what extent do badgers contribute to the transmission of tuberculosis among cattle?' contains the assumption that badgers do contribute to it. A more open-minded form of this question might be 'What are the major sources of tuberculosis infection in cattle, and are badgers one of these?' These contrasting questions represent different models to be tested. Model error also encompasses analytical error. Different analysts provided with exactly the same data can often come up with different conclusions.[6]

Finally, *implementation* or *interpretation* error is, in my view, where the most serious problems arise because this is the point at which knowledge is used, possibly as 'evidence'. All the other forms of error stack up to create a pretty fuzzy picture of reality, but if we decide in the end to ignore the fuzziness and zero in on what we want to believe from the unfolding story, then we are in danger of drawing the wrong conclusions about how the world works. Implementation error happens when those using knowledge do not appreciate or ignore the underlying error structure of evidence. The scientific advisers to the UK government during the COVID-19 pandemic worked hard to contain this problem. Others in the politics factory can be a lot less diligent, meaning that over-claiming about evidence can be the norm, and this is an increasing trend even among scientists.[7]

Unsurprisingly, all these opportunities to build error into scientific investigations mean that *evidence*, which sits at the top of the knowledge pyramid, can often exist on a pretty shaky basis. It is why, in the note I quoted above, I thought that most of the evidence presented by the public in a 'call for evidence' would likely constitute 'an incoming avalanche of unverified drivel'.

It is also important, however, to realize that rigorous application of the evidence pyramid is the best process we have to find out about the state and dynamics of the real world, and it is much more thorough than either fervent belief or adversarial debate, where baseless rhetoric and sophistry can creep in. It becomes essential to understand the process of evidence generation in order to be able to use the evidence which emerges, but this is mostly overlooked.

An illustration of this comes from a problem I encountered created by sober reports about declines in biodiversity, often regurgitated with little critical appraisal by journalists. 'Half of UK birds in decline' would be a common kind of headline, but it was my job to look at these kinds of tropes and decide how reliable they were. Often they turned out to be examples of error accumulating up the evidence pyramid topped off by interpretations driven by prejudice. In the case of trends in biodiversity, the evidence mostly came from data taken as *measurements* with quite high inbuilt measurement error. These were mostly a small and biased *sample* from a very large landscape and with only a fairly coarse understanding of the *processes* which led to the distribution of species across those landscapes. The measurements were then turned into information about the bigger picture of change in populations of a very select range of species by fitting those data to a pre-determined *model*, the outputs from which were then subject to *interpretations*, often by people who were preconditioned to see what they wanted. This was like taking a few fuzzy pixels from a complicated picture and using these to decide what was in the picture, often in quite a lot of detail. The picture developed by this method then tended to show consistent patterns of some populations of creatures declining and some increasing through time. As the news headlines said, this was interpreted to show that an alarming proportion of species were in decline.

But how true was this conclusion? The first problem is that the true fuzziness of the picture is mostly underestimated (or often not estimated at all); the equivalent of a map with no scale. The second is that improvements in methods and increasing data make more recent estimates more reliable than historical estimates. Fuzziness, therefore, increases the further back in time one goes, making trends a lot less certain than is usually admitted. The third problem is that the estimates made through time are mostly not statistically independent because they have often been made by going back to the same places year-on-year and they use the same methodology with the same errors in measurement and sampling. In some surveys, those places surveyed were originally chosen because they were places where birds or insects tended to be found in greatest abundance (such as nature reserves), which can precondition the resulting trends to show declines rather than increases in abundance. This happens because, in a fluctuating world, places where there was high initial abundance are more likely to show declines than increases. The fourth problem is that if we were to question, in the manner of an *impartial spectator* or a rational sceptic, what pattern of changes in abundance we could expect to see, we may conclude that in any fluctuating environment we could expect that at any moment in time about half of all animal populations would be declining while the other half would be increasing. So, given all this, is it right to interpret an observation of half of species being in decline at any time as anything other than what one might normally expect to see?

Understanding the process involved in accumulating error has certainly always made me feel humble about the extent to which I have a grip on reality, and I often marvel at those people who express high certainty in their interpretation of what is going on around them. In this specific case about trends in the abundance of species, it potentially has important implications for our interpretation of 'the biodiversity crisis'. That there is a crisis caused by the loss of biodiversity is very likely. Inductive reasoning based on the observation of the continual expansion of land use for agriculture and urban development, as well as the diffuse poisoning of land with chemicals, means we should expect biodiversity to be in decline and maybe even in crisis. On this basis alone, it may indeed be prudent to proceed on the assumption that there is a crisis, but we need to be very careful not to get caught up in manufacturing evidence through selective deductive reasoning, if only because it opens the whole proposition up to criticism by charlatans.

The proposition that climate is changing as a result of human activities is the classical example of this process where the charlatans got the upper hand for a while. A vast set of highly technical inductive analyses conclude that climate change is almost a certainty, and yet many people will not accept this until they are presented with direct, deductive evidence. They need to see (and probably directly experience) the damage before they will agree to act. This, in my view, is like expecting somebody to test the proposition that there is no traffic passing on a road by repeatedly stepping out into the middle of the road to see if they get run down. This is the subjective by-pass expressing itself in its rawest and most brutal form.

Rather than using inductive processes to guide their thinking and decisions, people who clearly sit inside the politics factory seem to have a need for deductive evidence before they can act. I suspect this is yet another artefact of the mis-appreciation of the difference between legalistic and scientific definitions of evidence. Evidence is then procured intentionally to reinforce what is already known.

In one example which troubled me, this process led to a minor industry developing within the scientific community to demonstrate the effects of neonicotinoid pesticides. It created large numbers of studies which were concocted intentionally to demonstrate effects rather than to honestly seek to test the opposite (the null hypothesis). What many scientists who get bound up in the politics factory tend to forget is that a study which sets out determinedly to demonstrate a null effect (like no impact of neonicotinoid pesticides) but which still finds effects is much more convincing than the weak-minded alternatives manufactured to support rather than to challenge the original proposition, in this case that neonicotinoids are harmful.

The combination of *model* and *implementation* error presents people in the politics factory with huge opportunities for the subjectification of evidence. 'What is the evidence?' is the question which many civil servants and ministers often ask when faced by some proposition. Mostly, however, they are seeking legally based rather than scientifically based evidence. My first question in response would be 'Whose evidence?', because it is very easy for personal preferences to creep in at all stages of the evidence generation process. This is why the evidence-generating machinery within the politics factory is tuned to come up with the results which are most politically acceptable. Anything else would be inconvenient.

Asking 'What is the evidence?' may be a genuine effort to apply objective reasoning, but it has become a kind of alchemy for re-framing subjectivity as objectivity. It throws people inside the politics factory into a complex knot of confusion which can then resurface as gobbledygook masquerading in the clothing of authoritative science-based knowledge. It is what drives governments to create the knowledge they want rather than the knowledge they need. Many people inside the politics factory are up to the same game, including members of some scientific communities.[8] One campaign for betterment suggested that 'the public – and officials and experts – should be able to see the evidence behind policy decisions', and 'only by seeing the evidence used can outside experts evaluate and add to it, and improve the evidence base for important decisions'. The suggestion was that 'we need an evidence standard'.[9] If only it were so simple. Evidence is not a commodity to be manufactured. When evidence itself has become corrupted by its procurement within the politics factory, these kinds of aspirations get re-cast as yet another political manoeuvre to control what evidence actually says. Scientific process already exists as a standard, but perhaps there is a need to hold people to account for how they apply it.

The challenge which needs to be addressed is that the notional requirement for objectivity in decision-making bestows power on the evidence providers and there are massive incentives to fix the evidence in ways which align it with the ideologies of those who generate the stuff. It is unsurprising, therefore, that the evidence gets fixed. It is the reason why many of the mid-20th-century think-ers in political science thought that an evidence culture could be toxic because power is shifted to those who look like they have technocratic know-how even when this is a deception.

The accumulation of fuzziness in evidence as it is created and the capacity for fixing the evidence paints a dystopian picture of how those within the politics factory rarely pin themselves down to reality. Even the army of fact-checkers which has grown up inside the politics factory, mainly in news organizations, is challenged to distinguish fact from fiction when the machine is so infected.

What can we believe? *Official statistics*[10] and science coming from trusted voices (who are evidentially and verifiably outside the politics factory) are about the only reliable sources, but the most reliable will always be those people who emphasize what they do not know. This mostly excludes governments, politicians, special interest groups, big business or commerce, self-defined experts, opinion-formers on social media and journalists. These are all operatives within the politics factory and rational scepticism is needed in order to combat them.

The note I sent to ministers and senior civil servants quoted at the start of this chapter emerged from recognition of the brazen ways in which evidence is corrupted. The corruption is there for us all to see, but we mostly pass it by because it is normalized. Journalists produce corrupt reports all the time, even in reputable news feeds, purporting to reflect public opinion based on interviews with very small sample sizes of people. There is really no pretence at respectability for those operating inside the politics factory. So prevalent is the problem that I got 'you're not being serious' looks when I complained. I saw these very people using *data* as *evidence*, imaging that one could jump from any level in the knowledge pyramid straight to *evidence* without bothering with all the steps in between. Mostly, these people were in a sea of confusion about the difference between legally and scientifically accepted definitions of evidence. Often, they just wanted to concoct a pre-cooked storyline from biased evidence, and if nobody came up with an adversarial response, then it became accepted. For example, if we want to know about demand for ice cream and we ask somebody at random if they like ice cream, their response is a data point. If they say 'no', this is not evidence that people do not like ice cream. (It may not even be evidence that the person asked does not like ice cream.) But that is the way many people in the politics factory use data as evidence. If somebody responds to a 'call for evidence' about ice cream by saying they don't like it, then that is taken as valid evidence that ice cream is not liked. Moreover, the only evidence one gets in these circumstances is from people who care about ice cream. The real position can only be illuminated when these responses are combined with a sufficiently large, unbiased sample of other data about preferences for ice cream from across the population of all relevant people.[11] This allows data to be combined into information which tells us something meaningful about people's general preferences. When it is then combined with other sources of information, such as the availability of ice cream, or the experience which people have had of ice cream, or the prevalence in the population of lactose intolerance, it starts to become knowledge. When used within a context of policies towards the regulation of ice cream, such as whether all ice cream should be made from pasteurized milk, it becomes evidence.

Left to ferment within the politics factory, evidence becomes a social construct,[12] a corruption of *quod erat demonstrandum*. Evidence in this form is about as deceptive and useless as a baseball on a golf course.

9

'What works' in the politics factory?

An offshoot of the corrupt evidence bandwagon in the politics factory is known as 'what works?' This is another poorly defined, ambiguous expression which has gained momentum as a way of thinking about problem-solving within the politics factory. It has been helped into this position by its use within parts of the scientific community, especially in the social sciences. When somebody from politics asks 'what works?', then they are asking the same basic question as 'what is the evidence' to support my proposition? Sometimes it is posed as a genuinely open question, but mostly it is not. It generates the same kind of pathologies as the evidence bandwagon but does so in a manner which lends a deceptive sense of thoroughness and objectivity to the task which makes it even more believable. Bertrand Russell put it this way, 'I have always found that the hypothesis of Santa Claus "works satisfactorily in the widest sense of the word"; therefore "Santa Claus exists" is true, although Santa Claus does not exist.'[1] The application of 'what works?' requires a thorough understanding of the proposition, and mostly this understanding is absent when 'what works?' is used within the politics factory.

Governments have invested in research capabilities, sometimes called *what works centres* or *networks*,[2] to address questions they have about whether proposed policies are likely to be successful. For example, one might want to know whether sending a text message to people one, two, four or eight days in advance of a hospital appointment influences the number who turn up. A 'what works?' study could then be carried out on a small subset of a population, for example within one hospital catchment, to assess the viability of such a scheme overall. This is a fair question and a fair thing to do. There is nothing wrong with the 'what works?' question in this context so long as the results of those studies are interpreted and used correctly, that is, in the same context as that used to generate the results and with the same attention to underlying causes of *error*, as discussed in the previous chapter. Problems arise as a result of ill-disciplined interpretation followed by application beyond the context of the original experiment. Sometimes this is reasonable, but only if the uncertainties associated with moving beyond the specific conditions of the original 'what works?' study are properly accounted for. What works for text messages in one hospital catchment cannot be assumed will work to the same extent in others. More seriously, this is why 'what works?' studies in the form of clinical trials for drugs and therapies which focus mainly on white Caucasian participants cannot necessarily be extrapolated to black or Asian communities. It partly explains the failure of COVID-19 vaccines to reach ethnic minority populations.

In the words of the UK government, their 'What Works Network uses *evidence* to improve design and delivery of public services' (my emphasis).[3] Underlying this is an idea which is not robust: the idea that if something works in a particular time or context, then it is also true in other contexts. Recall that *evidence* is knowledge used in a specific context. *What works is true* is not true in all instances, but once the evidence has been generated there is a tendency to assume that it is. It is the territory of psychics, faith-healers, and believers in UFOs and transubstantiation.

As I explained in chapter 3 one of my most trying experiences as a science adviser in government concerned the culling of badgers to control bovine TB. Much of the perversity about this policy emerged because it had been built around the results of a 'what works?' experiment,[4] which was fine as far as it went, but the results had lots of uncertainty and caveats attached to them which were often ignored by those operating within the politics factory – including both those who opposed and those who supported the culling. The assumption underlying the policy was that if the conditions of the 'what works?' experiment were replicated at scale, the outcome would be similar. The problem was that the results were likely to be specific to the time and location of the original 'what works?' experiment, so it was a stretch to imagine that the same outcomes were likely in different circumstances. This kind of problem arises because 'what works?' is often used in a legalistic sense as a true–false test of a proposition when it is actually a scientific test with uncertainty attached to the outcome. When people within the politics factory made the transition from science- to law-based interpretations, then this created a presumption that the result of the 'what works?' study of badgers was a general truth, something which, like Santa Claus, was almost certainly untrue.

Charles Sanders Peirce, one of the originators of the philosophical school of early 20th-century pragmatism, promoted the central idea that *what is true will work.*[5] He was saying that if we work hard to converge on truth, then we will also converge closer to what works. In Peirce's view, science generates generalities which apply broadly as a set of principles which can be applied in many different circumstances. These truly reflect reality and are therefore highly valued. Knowing whether traffic on the road keeps to the left or to the right would be an example of a general truth, and by knowing this we are able to build safe methods of using roads in very different circumstances. Many other things flow from this single piece of knowledge, so when research is focussed on finding these basic rules or truths, it is likely to result in a robust understanding of what is going on.

However, Peirce's friend William James reversed the logic of Peirce's statement to *what works is true*, creating a much more individualistic, perhaps even subjective, view of truth. Superficially, *what works is true* is a reasonable proposition, but in fact it is quite dangerous because the logical flow is opposite to that of Peirce's view. Instead of going from the general case to the particular case, it flows from the particular to the general, and this is the way 'what works?' is usually applied in the politics factory.

For example, just because I experience traffic on a road keeping to the left, this does not make this true in all circumstances. St Andrews, where I work, is not just home to a university but it is also 'the home of golf', making the small town a mecca for golfers from all over the world. This also makes it a place where people who come from countries which drive on the opposite side of the road are present in abundance. As a driver, one quickly realizes that what works for them in their homelands is not true in St Andrews as they step out onto the road in front of you without looking.

Peirce's form of *tough-minded* pragmatism would impose recognition of this difference. In contrast, James' form of *tender-minded* pragmatism would continue to assume that what is true for St Andrews works everywhere, and it is James' kind of pragmatism which is most often taken up and integrated into the politics factory.[6] It provides a kind of 'truth' which can be personalized and made to fit with ideology. It is an easy route to generating evidence to fit people's prejudices.

Many researchers are experienced at observing that what works is not always true. For example, there is an increasing understanding that assisted technologies invented to help with disabilities do not always reflect the success of laboratory trials when used in the field. What works in one circumstance is not necessarily an indicator of what works more generally. Thankfully, the idea of *what works is true* is rarely used in acute safety-critical situations such as licensing the operation of a new kind of aircraft or when approving the use of new drugs.

In the case of drugs, a much more sceptical process is used. This usually involves multiple layers of trials with appropriate statistical designs including *double-blind* methods, where participants and clinicians are both unaware of which treatment is being given. In fact, the tough-mindedness with which drugs are approved for use converges much more towards the *what is true will work* pragmatism of Peirce. This thoroughness is sometimes so overbearing it is even seen as anti-pragmatic[7] and so stringent that it adds greatly to the costs of developing new drugs, a factor which stands in the way of the development of new antibiotics to address the looming problem of antimicrobial resistance.

We need to find an appropriate balance between these two opposing approaches, and this is a valid point of debate. But the disaster which can ensue from the application of *what works is true* is exemplified by the procedures used to license pesticides.[8] Over the decades since pesticides have become an essential technology for producing cheap food, we have seen successive waves of new classes of pesticides coming on the market for them only to be withdrawn because of their damaging effects,[9] something which happens because of a tender-minded approach to investigating their effects. A pesticide starts to be tested in much the same ways as a drug. It is put through a range of tests on specific species to understand its effects and to search for side-effects. If it is found to pass those tests, it is provided with a certificate for use. Even if the conditions of use are specified, there is a tacit belief that the tests done on example species apply to all other components of the environment, a case of the flawed logic of *what works is true*. When licensed for use, drugs are normally rolled out under the rules of pharmacovigilance, which is an acknowledgement that even the most rigorous testing will not cover all of the real-world circumstances of drug use. There are rigorous systems, therefore, for reporting any adverse outcomes. Not so for pesticides, where tender-mindedness has been applied.

The result is that there is an assumption that pesticides which have passed small-scale tests can then be used at all scales. This was the problem I faced when dealing with neonicotinoid pesticides: at small scale and in specific test circumstances, they looked to have low toxicity to non-target species, but farmers then used these at enormous scales, meaning that whole landscapes were doused in these chemicals. They were being used well beyond the circumstances of any scientific studies which had been conducted to understand their effects. The integrated results of a plethora of studies (of very variable quality) suggested strongly that these chemicals were damaging ecological systems. Neonicotinoid concentrations were building up in soils, to, most likely, sterilize them of some forms of life, and they were being found in places well away from where they were being used. It was starting to create a picture of systemic damage like that seen from the use of organochlorine pesticides in the 1950s, the story which Rachel Carson told in her book *Silent Spring*.

People within the politics factory can be deviously bipolar when it comes to how they use the 'what works?'

narrative, depending on how much they think they can get away with diluting thoroughness and rigour. How drugs are approved is at one end of this bipolarity because when things go wrong, then people's lives are at risk and it is often easy to assign accountability for failures. Many other issues of equal importance to the health and welfare of people, such as the problem of how to minimize the impacts of poor air quality on health, sit at the other end, where it is common to find *what works is true* being put into practice. This is because the effects are diffuse and it is harder to lay blame. As a result of this difference in accountability, more than ten times the number of people die from poor air quality caused mainly by vehicle emissions than from road accidents, where accountability is easier to assign.

A further objection to the *what works is true* framing is that there may also be useful falsehoods, and believing that something is useful, or works, does not mean that it is true. As the quotation from Bertrand Russell at the beginning of this chapter points out, Santa Claus may be useful but he certainly is not true. Stories are often untrue. One can think of all sorts of equivalents in politics. Various rhetorical devices are used in efforts to help people to believe things which are not true. The advertising industry is especially adept at this. These ploys only *work* in the sense that they get people to do things other people want them to do often under false pretences and often against their own interests.

Useful falsehoods – like Santa Claus – are ubiquitous within the politics factory. When the UK government responded in 2010 to the so-called 'Aichi targets' for ending global biodiversity loss by producing a 'Natural Environment White Paper', it knew that there was no practical route by which these targets could be delivered. It took until 2015 for it openly to acknowledge this deception (along with the many other countries which had signed up). The same happened when the world breathed a collective sigh of relief when the Paris Agreement on climate change made it legally binding 'to limit the [global] temperature increase to 1.5°C above pre-industrial levels'. But this was a deception and the legal limit had been breached in less than a decade. The story is the same for the obesity crisis, food system transformation, the circular economy and many other contentious issues where the act of target-setting becomes the end point rather than the hard follow-up work of making any real difference. Some scientists, and many other intelligent people who should known better, fall into line with these falsehoods. What worked, in these cases in the context of international brokering or national target setting, was untrue when it was made to confront reality.

The Aichi Agreement, the Sustainable Development Goals, the Paris Agreement and most other conventions of this kind have a *what works is true* flavour to them involving a lack of realism about how what works socially and politically in one space and time frame (often involving a group of people in a meeting in some specific location coming to an agreement) translates to what really does work in all spaces and time frames. *What works is true* has all the hallmarks of sophistry or even deep dishonesty. It becomes a sophisticated way of lying.

Lack of rigour in 'what works?' therefore creates serious problems for humanity and the planet. Other common useful falsehoods arising from 'what works?' include money, because the value of money is certainly not a truth in the sense meant by *what is true works*. Unlike something tangible like raw materials, money *per se* only has notional value. The edifice of economic growth is equally a *what works is true* concept which struggles when it is connected to the natural world because common interpretations can imply infinite consumption of a limited natural resource.

William James introduced subjectivity to a form of pragmatism which Charles Sanders Peirce wanted to make as objective and tough-minded as possible and which Bertrand Russell hated. The illusionist James Randi advocated a *tough-minded* attitude when he famously said, 'Don't be too sure of yourself. No matter how smart or well educated you are, you can be deceived.'[10] He claimed that illusionists are the most honest people because they say they are going to fool you, and they do. They prove that what works is not always true, and if illusionists were to be dishonest, how would we know? Many folks within the politics factory are really dishonest illusionists because they deceive you without telling you. Then you are none the wiser. You end up making bad decisions and feeling unjustly dealt with even if you cannot always be sure why. *What works is true* is part of a dishonest illusion and it is part of the intentional ambiguity of politics. It is something which science, through the application of disciplined methods, tries very hard to counter.

'What works?' is also trendy because it feeds the needs that dishonest illusionists prey on. Popular ideas surface as a result and they amplify themselves through popular culture, often through social media. For example, there is an idea that there is wisdom in crowds,[11] implying that if we work collectively, we are more likely to come up with a robust answer to a problem. It is easy to see how this idea might be motivated by a political case for democracy over autocracy, but it is hugely dangerous and mostly wrong and is just a small step away from what Alexis de Tocqueville called the 'tyranny of the masses'. It is a classic *what works is true* illusion because it can only work under very specific conditions: where there is a free flow of information; individuals (even in the crowd) are acting in isolation from each other; they have enough basic knowledge to understand the proposition in question; and the laws of large numbers apply – equivalent to the law of averages. This makes the *wisdom of crowds* concept invalid in most circumstances.

'What works?', therefore, can be a genuine search for *what is true works*, but when science gets subsumed into the politics factory, it often ends up as a search for *what works is true* and this is often why policies end up in a mess. Tender-mindedness allows sloppiness to creep into the logical structure by which information and knowledge are generated from data. It allows the gradual infiltration of subjective discourse and illusion into objective reasoning. Many people who campaign for more use of science in government – for *evidence-based policy-making*, for example – actually implicitly campaign for tender-mindedness, sometimes out of ignorance of its existence and consequences but also sometimes because it is a way of promoting their own agendas by creating their own illusions. If writing this book does nothing else, I hope that it is read by those people who simply want more evidence-based policy or more research done to support the activities inside the politics factory and it makes them pause to think a lot more carefully about what they are wishing for. They need to understand what quality in science looks like and to promote this. We need to fix the tendency to reinforce the subjective by-pass by being not just tough-minded, but also sufficiently bloody-minded to re-design public policy around asking for evidence which is derived from rigorous scientific methods and which quantifies uncertainties.

10

Following the crowd

These days, one commonly asserted imperfection in the science–policy interface is that some so-called 'science' is imbued with policy preferences. Such science may be labelled as 'normative' and it is potentially an insidious kind of scientific corruption.

Robert T. Lackey[1]

Not all science is generated by people who honestly sit outside the politics factory. So, how can we identify research which is designed to create illusions? One of the most troubling aspects of the subversion of evidence is that many researchers are either oblivious to the problem or actively complicit in the process. Even some of the most august scientific institutions – the various national academies representing the pinnacle of science in many nations – sometimes have the capacity to reflect national interests rather than objective analysis in their deliberations on particular subjects.[2]

Scientific activity varies in its susceptibility to descending the chute of the subjective by-pass to land as a construct of people's prejudices. Science which is susceptible to political influence – where there is high uncertainty associated with problems, an urgent demand for the results and strong values-based judgements about what constitutes the right kind of result – has been called 'post-normal science' by the sociologist Jerome Ravetz.[3] Scientists often struggle when confronted by heavily socially modulated demands of this type because it is hard not to get drawn into the politics factory when conducting research to address these post-normal problems. The contemporary philosopher Michael Strevens has contrasted the resulting drift towards irrationality in the scientific process, what he calls *radical subjectivism*, with the call from mid-20th-century philosophers of science like Karl Popper and Thomas Kuhn for strictness in method.[4] This strict method involves defining and testing hypotheses in a process designed to converge on truth by gradually whittling away at uncertainty. Rather than diving into a problem and searching for *the* answer or *the* truth, which would be the *radical subjectivist* method, the stricter method converges cautiously on truth and probably acknowledges that it never quite gets there. In Strevens' view, scientists struggle to conform to this kind of methodological discipline because of the demand from people within the politics factory for absolute truth – legalistic truth – rather than relative truth, defined in terms of probabilities. Only priests, shamans, charlatans, lawyers and politicians deal in absolute certainties. Scientists often give in to the pressures around them and dilute the quality of the science they produce as a result. *Radical subjectivism*, therefore, is something we need to work hard to avoid.

Scientists can be as guilty as most other members of society of following the crowd, what Francis Bacon referred to as *idola tribus* (idols of the tribe),[5] by falling for fallacies like the *wisdom of crowds* or being susceptible to the *tyranny of the masses*. The challenge for scientists is that this is also often equivalent to following the money, and it is people in the politics factory who tend to hold the purse strings. Strong political interest creates a vicious cycle because it reinforces embedded biases which in turn entrain researchers to produce even more results to support those biases. Science becomes normalized or channelled to producing the results that people seek. It becomes *normative*.

Normative research represents scientific research done specifically to confirm current preferences. In my own experience, normative processes were especially prevalent in the research carried out on badgers, declines in biodiversity, the environmental effects of pollutants like endocrine disruptors, pesticides and plastics, air quality, ocean acidification, global warming, food poverty and the effects of diet on health, but they are prevalent in many other fields too, especially in areas or research which have immediate interest to people like human health or poverty. They were on display during the COVID-19 crisis: an effort to summarize all the information from research on the effects of 'test, trace, isolate', also known as TTI, during the pandemic identified twenty-five studies of which only one was carried out to the highest methodological standard, but even that was assessed to have 'some concerns' with regard to bias. Of the remainder, six were assessed to have 'moderate risk of bias' and the remaining eighteen were at 'serious risk of bias'.[6] Bias of itself does not signal normative research because some studies are just inherently difficult to carry out in ways which are unbiased, but there were doubtless strong normative components in this research. In agriculture, normative research is focussed on reinforcing an old and tired system of producing food which is both very inefficient and environmentally damaging. All these are subjects which are hotly debated within the politics factory or where there are strong vested interests and strong opposing views.

Public health science is especially stacked with normative research. The high demand for results encourages a flawed *what works is true* foundation of logic for many studies. Here, the *randomized control trial* is the much-abused gold standard, a method in which participants in trials are allocated at random to different treatment groups. Often, however, it is impossible for participants and clinicians to be blinded to the treatment and the participants are almost never a random sample from the population they purport to represent. The risks of bias in these cases is vast. A minor industry produces small-scale studies designed specifically to extol the health benefits of specific diets or exercise regimes and most of these are next to useless. In fact, it takes enormous effort and skill to design and execute robust epidemiological research in public health.

Normative research tends to be much less prevalent in fields like weather and flood forecasting or the measurement and dynamics of the state of natural resources like forests, fish stocks or oil and mineral reserves. These are all fields where there are more direct measurements which make science more certain. In materials science, a new material either does what it says on the tin or it does not. It is mostly obvious when it fails. Normative research is even less prevalent in fields which have little or no immediate effect on the interests of people, like particle physics, astronomy or quantum chemistry (although nobody should imagine that these fields are free from the effects of wishful thinking on the part of researchers).

Normative research is intimately associated with the loss of integrity in research and those carrying it out. It is infested with an array of bad behaviours which have names like *confirmation bias*: the temptation to go with the crowd and to confirm a previous study rather than to challenge it by generating new results from a different angle. Researchers whose only experience comes from working in particular fields, something called *normal* science by the scientific philosopher Thomas Kuhn, often have a problem with this kind of bias.[7] *Normal* science is often supported by *hypothesis myopia*, which is the tendency to collect evidence to support a hypothesis rather than rigorously looking for evidence against it. After all, if you believe in something, why not look for the evidence to support your belief? It was the philosopher Karl Popper who explained that scientists need to do precisely the opposite.[8] The pathology of chasing beliefs manifests as *asymmetric attention* involving rigorously scrutinizing or supporting some results more than others. This is sometimes also referred to as *cherry picking*, which is a more sinister descriptor because it suggests some of the behaviour is intentional and not simply inadvertent or culturally embedded. There is also *HARKing*, which is the presentation of *h*ypotheses *a*fter the *r*esults are *k*nown;[9] and *just-so story-telling*, involving the *post hoc* rendition of beguiling and plausible explanations for phenomena. Sometimes also known as *adaptive story-telling*, this behaviour is named after the 'just-so' stories of Rudyard Kipling and, as Stephen Jay Gould highlighted, is a tendency of much evolutionary biology.[10] This is a story which tends to tell the recipient what they want to hear and which confirms existing biases. Natural history films often reinforce these kinds of stories in the public mind. Environmental lobby groups use 'just-so' stories to aid their rhetoric about environmental doom, but so do their opponents. Many comforting stories about how new technologies will come to our rescue to deliver us from the harms of climate change also have a strong element of 'just-so' about them.[11] Concocting a story to fit observations is like being witness to an apparent miracle and then building the rationale for the observation by invoking the hand of God.

However, I do not think scientists instinctively want to behave in these ways. Instead, it is contact with the politics factory which incentivizes these behaviours. Evolutionary biologists are likely driven to *adaptive story-telling* by their feeling that there is a need to counter the hocus-pocus coming from anti-evolution groups. Actors within the politics factory, from government servants, politicians to corporate companies and pressure groups, want to skew the results of research to promote their own vested interests and scientists are subject to a barrage of influencing processes. In addition, scientists have their own internal conflicts to iron out in the form of the *scientifico-political predicament*, and some fail to realize this. They, themselves, become actors within the politics factory but deceptively proclaim their 'independence'. In seeking to be socially normative, scientists can betray their duty to be anti-normative.

Some people have argued that some sciences, such as conservation biology,[12] are intentionally normative. This could also include research, for example, on the relation between smoking tobacco and lung cancer[13] because it is searching for a specific effect. By this definition, any research with a purpose is normative, and so this could explicitly characterize an immense amount of science. Some sciences are even known as 'the normative sciences' (such as aesthetics, ethics and logic within the field of philosophy) because they investigate the normativity itself. But this is a very specific use of the idea of normative science and the danger is that anything with a sociological dimension is deemed normative.

It is essential to distinguish between this definition of normative when it is focussed on the *subject* of research as opposed to the *method* of research used to converge on truth. It is perfectly logical to study the norms of behaviour or to do research with a purpose, but it is *how* the research is conducted which determines the robustness of the answer. Tough-mindedness and rational scepticism are the hallmarks of research which combats normative tendencies.

In conservation biology, studies which are normative because they are focussed on providing answers to specific questions need to be embedded within a wider context of theory, such as ecological theory. These are rigorous structures which guard to some extent, although not wholly, against normative behaviours among those doing the research. They encourage the testing of ideas against null hypotheses rather than carrying out aimless searches for apparently interesting observations. The confusion between normative subjects and normative methods creates an excuse for diving down the subjective by-pass. It is only one small step from declaring that a whole field of science, like conservation biology, is normative to finding an excuse to use normative methods to such an extent that it is reinforcing preconceptions rather than addressing open, unbiased questions.

Normative research is the same as trying to write scientific papers backwards – start with the conclusion and then concoct the story to fit. I suspect this is why whole fields of science which are generally normative lose quality and credibility.

This is what Lackey, quoted at the top of this chapter, was really referring to. As he suggests, normative science has a lot of the qualities of 'corrupt' research. Only studies which take a tough-minded, sceptical stance to test the fallibility of a proposition are truly valid. Researchers need to determinedly set out to prove that a proposition like 'there is a decline in pollinating insects'[14] is wrong. Only by questioning the proposition and doubting the answers from research seeking evidence of no decline will we eventually convince the sceptical, tough-minded pragmatist that insects and pollinators really are in decline. This contrasts with the *normative* approach, which is to construct studies which are more likely to find declines, an approach which I found many people in the wide amateur diaspora which accompanies conservation science simply do not understand. The resulting loss of credibility of science in certain fields can also be very damaging. Climate scientists were criticised for some normative behaviours in the early stages of climate change research, which handed excuses to climate change deniers to undermine their research.[15]

Normative research can be hard to spot, however. For example, I once had to advise government ministers about a particular controversy where there were only two pieces of research available to support decisions. Both studies came to diametrically opposite conclusions and both were published in reputable scientific journals which applied rigorous quality control. However, the authors of one paper came from an institution with close ties to the industry which had a vested interest, whereas the authors of the other paper were from an environmental pressure group which had the opposite interest. It was clear that one or the other study, and most likely both of them, were examples of research which was produced to satisfy the prejudices of the communities the researchers came from. They had both been skilfully, and most probably unconsciously, constructed to come to a specific pre-arranged conclusion. They were both convincing illusions.

Sometimes clues to the presence of normative research can appear in the first few paragraphs or the conclusions of a scientific paper when the authors make an assertion, such as that plastic pollution is a burgeoning threat to the sustainability of our planet. This assertion is then supported by referencing secondary sources, including scientific papers or reports which were not themselves describing research and which either made the same unsupported assertion or stated that this was likely to be a problem. It is not the case that the more times an unsupported proposition is repeated, then the more likely it is to be true. What should have been a hypothesis, a proposition, had become an assertion. Rarely will research in a paper of this type ever overturn the assertions made, but if it was to be useful, it would do exactly that. In another example, I saw a supposedly neutral research funding agency publishing a call for research proposals to study *the* decline in pollinators. Rather than making an open-minded proposition, the agency was making an assumption about the certainty of a decline in pollinators when, at the time, there was no such certainty. These examples are the signature of a drift towards a normative justification for research. Normative tendencies, tendencies to go with the crowd, can close our minds to alternative explanations. This can have damaging consequences,[16] but the incentives for researchers to cook the books can be overwhelming when working in the domain of post-normal science.[17]

In response to these challenges, the contemporary philosopher of science Helen Longino has suggested that objectivity is eventually distilled out of the collective scientific enterprise and that the more diversity which exists within this enterprise the better.[18] Her argument supports feminist philosophy because of the diversity brought by the greater participation of women in science, but it would apply equally to other axes of diversity. In theory, the greater the diversity, the less the chances are of normative bias, but my experience suggests that when everybody is being entrained by politics, diversity makes little difference. Research integrity in politically controlled environments like central government departments is really an oxymoron and nothing much can be done about this other than to extricate research from those environments as much as possible.

Rather than seeking more diversity, an alternative response can be to ask for more research to be done, but this can just make things worse.[19] By way of an allegory, imagine if science is attempting to decipher the information within a digital image composed of pixels. If we start from a position of ignorance with all the pixels scrambled in random order, then the picture would be made up of all the same components as were in the original but it would contain very little information. The image would be unrecognizable. Gradually reordering the pixels back towards the original is like the researcher trying to build knowledge of the contents of the image by adding in information. Early research in this process will make most progress because the easiest parts of the problem will be solved first, but thereafter there is a law of diminishing returns as it becomes harder to add new information based on a standard amount of research effort. At best, this means that doing more research does not necessarily allow us to sharpen the image in the picture beyond a certain point. But could additional research actually result in a *decline* in information by creating a false impression of the image in the picture?[20]

I suggest this is possible, even likely, and it is caused by the addition of normative research. The reason is that as more researchers pile in to address the problem of deciphering the image, they add their own information to the problem. Each researcher leaves traces of their own prejudices behind within the picture. Their information about their preferences for what the picture might ultimately show becomes part of the picture itself, and other researchers trying to decipher the picture in future have no way of distinguishing between this false, 'zombie'[21] information and true information. As false facts become canonized,[22] the emerging picture becomes something different from the real picture. This is why the large language models used in artificial intelligence start to decline in their accuracy – they become polluted by their own inaccuracies, a process of positive feedback resulting in their own self-destruction. They become increasingly buried in their own excrement, but sometimes this also happens to whole fields of science. It is also why I believe that it is being optimistic to suggest that diversity of researchers is likely to improve outcomes, or, indeed, that simply asking for more research to be done to solve post-normal problems is helpful. More diversity arguably adds more diversity of prejudice. Only if the research is done to increasingly rigorous standards, to root out normative tendencies, is more research likely to be helpful.

Whole fields of science can get caught up in this problem. In forensic science, this has become known as the *bias snowball* or the *bias cascade effect*.[23] The cascade effect happens when irrelevant contextual information, such as might be collected at a crime scene, gets cognitively combined with strictly objective evidence. This might happen, for example, if the same people who sweep the crime scene for forensic evidence are also involved in subsequent analysis and report writing. This gives them the opportunity to build their own biases into analysis and reporting.[24] In forensics, it is possible to introduce strict, tough-minded pragmatic procedures which help to eliminate these effects, but in many scientific investigations there is relatively little appreciation of these kinds of issues, let alone procedural intervention to reduce their impact. Few scientists will ever be cross-examined by a defence counsel on their methods and procedures, but maybe that is what is needed.

I have found myself in the position of the defence counsel when it came to subjects like the effects of plastics on ocean life, the effects of endocrine disrupting chemicals on organisms, the transmission of bovine TB or the effects of pesticides on bees. As a research scientist, I was perplexed that the picture in many of these fields was becoming fuzzier as more research was being done. In broad terms, I saw that this happened because science was being increasingly deployed as an instrument of ideology.

Normative research is wasteful (and often harmful) because it promotes decision-making which, at best, has a false level of certainty and, at worst, might be wrong. It diverts money and talent to nugatory activities, so it incurs opportunity costs as well. It may even contribute to the apparent decline in recent decades in the level of disruptive innovation in science because by definition normative research is not disruptive.[25] Nobody has estimated how much of research is normative – and therefore how much resource is wasted – but I suspect it could be considerable. The anthropologist David Graeber reckoned that about one-third of jobs across society may be 'unnecessary, or even pernicious'.[26] Graeber was not thinking specifically about jobs in scientific research, but some research could be an unnecessary or even a pernicious activity and it would be reasonable to classify normative research in this way. Like Graeber's view that there are 'bullshit jobs' which we need to eliminate, perhaps we also need to work harder to eliminate 'bullshit research'.

One way to do this is to look most closely at the social chemistry of fields of research and to apply greater scepticism to science which comes from fields which conform to Ravetz's post-normal definition. When syntheses of research are conducted, as Longino suggests is the way forward, then less weight should be applied to studies coming from those fields.

Instinctively, this is what I think I was often doing as a Chief Scientific Adviser. I was instinctively applying scepticism, and a higher level of discounting, across the board in fields where I knew there was high political interest and the possibility of political interference from all sorts of players operating on the politics factory floor.[27] But for me, ultimately, the existence of normativity in scientific research has a much more sinister side than simply getting things wrong and living in the world of hocus-pocus belief rather than grounded knowledge. This was summarized by Hannah Arendt's *Eichmann in Jerusalem: A Report on the Banality of Evil*,[28] where she claimed that in totalitarian settings evil is normal. It becomes an embedded part of the internal structure of society through its laws and customs and is accepted by its participants.

There are obvious parallels between this and the processes across society which actively generate and promote falsehoods. Adolf Eichmann's defence was that he was a cog in the machinery of the Third Reich, and I suspect, if challenged, most researchers would put up a similar kind of defence about their role in society. By being drawn into normative modes of working, scientists certainly amplify banality and, even if most do not perpetrate the kind of evil which Eichmann was responsible for, they promote injustice. From tobacco science to climate and vaccine denial, these injustices rank with the Holocaust for their global impact. Normative research is not just 'corrupt', it is evil.

11

Trust in experts?

... people have had enough of experts from organizations
with acronyms saying they know what is best and getting it
consistently wrong.

Michael Gove[1]

Given all that I have said about the problems created by
the subjective by-pass, it is little wonder that people can
become confused by how evidence is presented to them.
The senior UK Cabinet Minister Michael Gove famously
took 'experts' to task when he was railing against a values-
based (and universally dire) set of economic predictions
about the departure of the UK from the European Union.
On the one hand, his view could be seen as populist rhetoric designed to undermine objective analysis which was
saying something he did not like. On the other hand, was
he an astute observer of the corruption of science? Were
experts being disingenuous, or deceptively political, so did
they merit being taken to task? Scientists certainly saw his
response as an attack on their integrity, but there is no
doubt that some sciences can give an impression of false
precision and are not as honest as they might be about the

uncertainties there are in their analyses, and they do this for political ends. Economics in particular can present as a quantitative art form creating an impression of objective analysis which simply does not exist, and it was mostly economics that Gove was talking about.

Economics has garnered special powers inside the politics factory. This is because of the tacit claim, often made tongue in cheek by economists themselves, that economics has the power to measure or quantify the unmeasurable, to make the intangible tangible or to quantify something as quixotic as 'value'. The real problem is that people inside the politics factory swallow these claims, thus lending economics the power to represent apparent reality in ways which are hard to justify in any objective sense and which, in the end, likely lead to injustice.

Procedurally, policies involving everything from the building of major infrastructure to the provision of milk to growing children in school cannot be implemented unless they are given the blessing of an analysis done by economists which suggests the policies are likely to create more benefit than cost. In theory, this makes a lot of sense, but the alchemy of economics involves doing things with data to create information which most other scientists would eschew. Economists might argue that their analyses are an exercise in realpolitik and are always better than the alternative, which is knowledge-free, belief-based and ideological decision-making. Certainly, in the event that a decision needs to be made, it is much better to make the decision with the benefits of the knowledge we have, but it is important to be open about the uncertainties in the results from economic calculations. To do other than this is to acquire power by deceit.

For example, when the options for improving air quality in cities were being investigated, I asked for an assessment of uncertainty around the economic net benefit of different policy options, but I was informed that they had not been calculated. The reason was that the estimates of uncertainty would be so large as to make the whole exercise seem pointless. This was the same as saying the numbers being presented were close to fiction. I saw the same problem in the economic assessments of food security, and in other fields too. This left me feeling profoundly worried that the supposed 'objectivity' of economic analysis was simply a shroud for a neo-liberal form of politics supported by a particular brand of economic thinking.[2] It seemed like black-box models, often controlled by a few junior apparatchiks, were determining government policy.

Besides turning science into a legitimate target for people like Gove, this shifts power into the hands of those doing the analyses. Bias can be introduced into this kind of evidence subliminally, but there was a distinct possibility that economists were being more than subliminal in their biases when it came to the arguments about the departure of the UK from the European Union. External scrutiny of their workings was mostly not tolerated. When addressing a despondent group of government economists immediately after the UK had voted to leave the European Union, I accepted that the vote was most probably an act of gratuitous self-harm, but I also said it was a sobering message for people like them. It told us a lot about how membership of the EU – a product of the very economic thinking they mostly adhered to – had largely failed to convince people that it was working for them.

The economic principle of free exchange of goods, capital and labour which underlies an economic union is all very well in theory, but these economists had fallen for the *ecological fallacy*: the treating of individuals who have free will as just another commodity, a number in a computer, and with an expectation that they will behave and believe in standardized and predictable ways. In this case, their

assumptions had been rejected. A sizeable proportion of
people did not like the homogenization of markets, the loss
of cultural heterogeneity and being treated in the labour
market like a commodity. There was an upper limit to
which people would accept technocratic solutions. When
mixed with an unhealthy dose of xenophobia, this became
the vote to leave the EU. It was a warning of sorts to those
who thought that technocracy could guide decision-making
on its own. Like good scientists, they needed to accept their
experiment had failed and go back to the drawing board to
think again.

But I do not want to be critical of economists in particular because other experts make similar mistakes. The
story about genetically modified organisms (GMOs) has
a complex history but is fundamentally an issue of how
experts are viewed and whether they are trusted. For years,
geneticists and agricultural scientists have lectured people
about the advantages of GMOs without understanding that
it is questions like 'who benefits?' or 'what's in it for me?'
that matter.

In some places, like the US, food made from GMOs has
been broadly accepted but in Europe it is still not permitted.
Even today, I see experts making the case for GMOs to be
permitted based on arguments mainly about the benefits
this will bring for the economy, for farmers and for them
as research scientists, because they operate under the tacit
assumption that the public believes research scientists
would only ever work to enhance public good. But when
all their hyperbole is removed, the public benefits are often
scant, poorly explained and mostly implicit. People are left
to work these out for themselves and they struggle. Many
people who become critics of GMOs are actually scared
because they look like yet another thinly veiled attempt to
dupe people into accepting a technology which is really just
going to benefit a few, not least those who run corporate

America. In places like the US where the flow of private benefits[3] is perhaps more accepted than in Europe, GMOs have crept over the threshold of public acceptability. Much of the trouble is caused by experts being unable to set aside their own self-interest and failing to create a trusted profile with the public.

The 'what's in it for me?' question also affects attitudes to vaccines, and the narratives coming from experts often forget to take this into account. Vaccination is one of the most powerful methods we have to control infectious diseases and yet it is under threat because of a lack of trust. It is often only effective if a very large proportion of a population is vaccinated, so there is a need to appeal to people's social conscience at least as much as to the 'what's in it for me?' question. But often that very question is not addressed with sufficient vigour. Understandably, experts in epidemiology have their eye on the necessity to get as many vulnerable people vaccinated as possible, but there needs to be a tangible benefit for individuals as well. Anti-vaxxers may be selfish sociophobes, but experts need to become much better at explaining and demonstrating how the benefits from vaccination flow.

Scientific experts can very easily get captured by powerful elements within the politics factory and become a mouthpiece for vested interests, as happened with Norman Borlaug within the context of the 'Green Revolution'. Often, the fear among the public can be that experts are not all they seem to be. Trust is an essential ingredient for the effective uptake of science, but once lost it may never return.

Arrogant, entitled and elitist are all adjectives I have heard applied to scientists, and none of these engender trust. Barack Obama referred to some lawyers as 'highly credentialed, high-IQ morons',[4] a description which also fits a fair number of scientists as well. Every profession has its morons and some also rise to the top. Sniping about science and its fitness to participate in government has turned into a minor industry for some in the press, but sometimes the science community walks into their poisonous narratives with breath-taking naïvety and GMOs and vaccines are classic examples. As scientist and philosopher Edward Wilson observed, 'so many accomplished scientists are narrow, foolish people'.[5]

The relationship between expertise, trust and the agency to control agendas is a rich area for exploitation within the politics factory. Scientists often innocently get caught up in the deceptions being played out or they can use the trust in their position to become part of political gaming. Some take advantage of their power over the public as trusted personalities, and even some totems of trust, like Nobel Prizes, can be undermined. Michael Levitt, a structural chemist and Nobel laureate, fuelled dissent about responses to coronavirus in 2020 by deriding 'lockdowns'. Even if they were an unpleasant and controversial measure, we would have expected somebody of his intellect to have understood the dilemmas involved. Linus Pauling, a double Nobel laureate, lost his way when he incomprehensibly started to promote vitamin C as a cure for almost everything; Jim Watson, the co-discoverer of the double helix of DNA, was deeply racist; William Shockley, who co-invented the transistor, promoted eugenics; Norman Borlaug praised the use of DDT in agriculture; and Kary Mullis, who developed the polymerase chain reaction for amplifying DNA fragments, denied that AIDS was caused by HIV.

Clearly, in these cases, the proxy of quality in the form of a Nobel Prize was not a good guide to the kind of quality which made these people's opinions useful or correct. But the deception of totemic status means there can be exaggerated respect for those who are *experts*[6] – once recognized, it becomes easier and easier to gain further recognition. It seems that scientists themselves are prone to creating this effect because they show a bias towards the famous even in their own research.[7] This collective raising of certain individuals onto a pedestal is probably one reason why it becomes so difficult to rid science of racism and sexism.

There are those, however, who just dislike experts, either because any group which professes to have power over knowledge will be looked on with suspicion, or because experts often do not reinforce prejudiced views. Donald Trump called experts 'terrible'. As much as he could, he selected the experts who would tell him what he wanted to hear, and this is a common failing within many political systems, even in those which are supposed to be some of the least corrupt.

Bashing experts is a common theme within populist politics. After making a public comment during the COVID-19 pandemic,[8] I got sent the following by one disgruntled reader: 'Frankly I'm a bit fed up with reading and hearing about how wonderful our experts are.' I responded to this saying,

> The dilemma we face with 'experts' (a term I dislike) is that they sometimes tell us things we don't like to hear, but they are usually better informed than we are, so we are forced to face up to what they are saying, even though it can be counter to our instincts or values. We turn to experts all the time for help: dentists, joiners, plumbers, doctors, scientists, economists, engineers, etc. How would you like to have a tooth filled by a non-expert?[9] The best expert is one who knows how much she/he doesn't know and can transmit that through their advice. What others then do with that information is for them to answer for.

Even if we do not like them all the time, we need experts. The question is: how much should we trust them, and how do we know they have not been corrupted? We need a basis for judging how much they can be trusted. After all, just about anybody can call themselves an expert, and there are few signals which convey the quality of the opinions many experts provide, especially in the sciences.

Some scientists who have looked at this subject suggest we should 'use experts wisely'[10] by relying on multiple rather than individual expert opinions, and ideally the multiple opinions should come from people with diverse views and backgrounds; being careful about choosing whom to listen to by knowing something of their background and their qualifications while not over-valuing experience and repu-tation; valuing those experts who are humble rather than assertive; and, if possible, testing the experts with things you already know the answer to. Experts often convene as groups, like science advisory committees, but they need to be reminded about the dangers of groupthink, factionalized views, bullying, egocentric bias, over-confidence, conserva-tism, risk aversion and confirmation bias. Drawing on his experience as a Chief Scientific Adviser, Bob May would have added declaration of interests to this list. He said, 'do everything openly (no consensus-forming behind closed doors)' and 'admit, indeed emphasise, uncertainties'.[11] I have seen some truly terrible behaviours in science advisory committees which were ignoring all these points. Honest feedback about these kinds of behaviours and the success of their advice is important.

Doing everything openly has its down sides, though. Some experts have a habit of playing to the gallery, which is not only irritating but also undermines consensus. Members of the UK's Scientific Advisory Group for Emergencies (SAGE) during the COVID-19 pandemic received a lot of attention from people with crackpot ideas about how to cure COVID-19; they were also threatened, sometimes with violence. Openness, therefore, creates other problems because not everybody is prepared to respect conventions of behaviour (let alone the law) when it comes to how they interact with experts, even those who are trying to do their best in difficult circumstances.

Experts in some professions get immediate feedback about how good their opinion has been, such as weather forecasters and intensive-care physicians. These experts are thought to make fewer mistakes because the feedbacks encourage greater care and allow faster learning. The feedback provided to the epidemiological modellers predicting the spread of COVID-19 made them progressively better modellers. But in many fields of science there is little by way of strong feedback of this type. It is impossible, for example, to verify a climate scientist's advice if it concerns the state of the climate in fifty years' time. Instead, climate scientists have had to find other ways of refining and verifying the models they use,[12] but not all science disciplines are so meticulous.

A rarely acknowledged conflict experienced by experts is that they can find it hard to be impartial. Smith's theoretical *impartial spectator* is a hard condition to attain, but economists and social scientists in particular are often like the eyeball that has to see itself. They are themselves structurally embedded in the systems they are observing and cannot disengage from the downstream effects of what they observe. There is a maxim in quantum physics that in order to measure something, then you have to change it; this also applies in many social systems, because the observer often cannot sit outside the system being observed, and the closer to the centre of policy-making that person sits, the greater his or her impact. It is hard to argue that a cosmologist's projection of the future has any influence on the evolution of the cosmos or even that a meteorologist's projection affects future weather. But it is difficult to externalize economists and social scientists from the very systems they are observing.

This co-variance of scientific prediction with social system dynamics can create confusion about what the predictions of experts can achieve. Projections of high death rates from models of COVID-19 often did not come to pass precisely because the projections themselves were rendered inaccurate by the reaction politicians had to those projections, something some politicians and journalists found hard to understand in the commentaries they gave after the fact. Their claim that the advice they received turned out to be wrong was because they reacted to the advice, which is precisely what was intended. It was they who made the prediction obsolete, and we are all better off as a result.

Despite scepticism about experts, a large majority of people think science is important, irrespective of their political leanings.[13] A sizeable minority think science carries risks but want it to focus on making the world a better place. Polling shows that scientists, the experts, are likely to be some of the most trusted professionals.[14] Even people who are religious appear to trust scientists more than their own religious leaders,[15] which perhaps suggests that when it comes to serious issues, even religious hocus-pocus takes a back seat in people's minds. In contrast, politicians are some of the least trusted people.

If scientists are some of the most trusted people in society, why are there not then demands for these people to run for political office? The reason is probably that it is the job which makes the person, and scientists would just transmute to become untrusted politicians. The only minister I worked with who was a trained scientist with a Ph.D. was indistinguishable from other politicians.

I also suspect we like to have untrustworthy people representing us. I once had a boss who could be completely unreasonable, but I found this tolerable mainly because I reasoned that he was also capable of being completely unreasonable on my behalf. So it may be with our politicians. We do not elect them on the basis of trustworthiness because we do not want politicians who are necessarily nice, honest people. Our politicians are convenient proxies for our own need to behave dishonestly while also avoiding accountability. This has a strong message for how to construct better ways of including science within the politics factory.[16] Just electing scientists to occupy the role of politicians – effectively transferring the expert into the politician – cannot be the answer. But neither is setting scientists up as a group of experts – technocrats – who gain power by controlling the information and knowledge which govern people's lives.

Perhaps the ultimate expression of the cult of the expert comes in the form of the 'scientific consensus'. This has commonly been used to combat disinformation on and doubt over climate change, but also in controversial areas like vaccination, the teaching of evolution in schools, the health and environmental impacts of meat-based diets and many other health-related, post-normal issues. But consensus-forming has some of the same pitfalls as expert groups and it has the same kind of problems as the illogical *wisdom of crowds* paradigm. Just because a lot of scientists – many of whom have little detailed knowledge of a subject – say something is so does not necessarily mean it *is* so. Methods have been developed for deriving consensus which are designed to minimise the effects of biases and focus on forming a clear and knowledge-based message.[17] Consensus-forming is a response to the rejection of knowledge by decision-making actors in the politics factory, but, if used unwisely, it could turn the relationship between experts and politicians into a spiral of distrust. It could become a tool by which one tribe within society wields power by telling other tribes how to live and behave. It may be an effective tool in some circumstances, but it needs to be used carefully and sparingly if it is not to be seen as a weapon in an arms race for power.

The problem for non-experts is how to make a judgement about whether an expert is an imposter or whether groups of experts coming to a consensus view are worth listening to. How are committees like SAGE to be trusted? Much work has been done in social psychology to understand this problem. Experts are knowledgeable about certain things but they are also citizens who bring their own values to debates. It is not easy to signal clearly when experts are moving from their role as champions of objective knowledge to their role as a citizen, and their right to hold an opinion. We need mechanisms to help people judge the integrity of experts. This is about recognizing quality and ensuring it is badged appropriately.

12

Redefining quality

I have argued so far that science has a pivotal role within politics but that quality in science rarely survives contact with politics. One reason for this is that generating high-quality science is not necessarily a pathway to engendering trust, and the result is that there are incentives to erode quality, to undermine the very basis of science as an ethical system of investigation. These incentives translate into systematized, invidious processes which subjectify or even fictionalize evidence.

Trust in science is essential if it is to have a role in future, but in the politics factory there are two sides to trust. There is the trust as seen by the politician, *subjective trust*, which asks whether you come from the same tribe and whether you hold compatible ideologies. It is the form of trust which asks the question 'Can you be trusted to act in my interests?'

Then there is *objective trust*, the trust we normally associate with the relative truthfulness of somebody which asks whether that person is truly striving for objectivity and will never gild the lily irrespective of your or their interests. Grudging respect exists for those who engender objective trust, but there is a strong preference for subjective trust within the politics factory. As a result, much of its machinery is geared to levering people who strive for objective trust – normally scientists – into the mould of creating subjective trust, a process which corrupts the pursuit of objectivity.

Therefore, when scientists appear to be, in Sheila Jasanoff's words, 'drunk with the mastery of pathbreaking scientific insights', and where 'experts too often underestimate the complexities of hybrid socio-technological systems',[1] something has gone wrong. One interpretation is that it is the same as the breath-taking naïvety shown by scientists when dealing with 'socio-technical systems' like genetically modified foods, but this is also possibly a view looking outwards from within the politics factory at people who intentionally stay outside. It is an implicit criticism from those who value objective trust but who would rather it was re-framed to align with their beliefs and ideologies.

A pessimistic view might be that the resulting corruption is getting progressively worse as the politics factory manufactures more and more imaginative ways of undermining robustness in science, effectively by tipping the assessment of trust and the quality of science towards a 'what's in it for me?' perspective. New kinds of scientific fraud are being discovered all the time, such as when a group of computer scientists stumbled on some odd phrases being used in some scientific publications. These were tracked down to reverse-translation software being used to cover up plagiarism. Subsequent analyses showed that there was widespread use of this method to cover up the operation of *paper mills*, which are sections of the politics factories which are specifically designed to produce entirely fictitious scientific publications.[2] Many reputable scientific publishers are having to develop sophisticated screening of scientific papers to weed out these kinds of frauds.[3]

These kinds of behaviours can get attributed to human frailties involving cognitive bias on the part of researchers,[4] but the more fundamental reason lies with how scientists are being entrained to think and work in particular ways. It is like a belligerent, controlling beast exists at the heart of the politics factory. The result is 'what works?', evidence, normative research and the cult of the expert. These kinds of developments threaten the survival of science itself. For all the discoveries of the activities of *paper mills*, how many unidentified pathologies are there actively undermining the value and quality of science? How, therefore, can we define and recognize quality in scientific outputs and make sure it is supported and that science survives?

Quality was the focus of Robert Pirsig's brilliant modern philosophical dialogue *Zen and the Art of Motorcycle Maintenance*.[5] His thesis was that there is something ethereal and intangible about quality. It has the characteristic of being undefinable but 'you know it when you see it'. Socrates called this *arete*.

I am always suspicious of fellow scientists who seem to be clear-minded about what is meant by excellence in science.[6] They will sometimes arrogantly make judgements about the comparative quality of research as if it is a contrast between black and white.[7] Often this reveals prejudice rather than a judgement about 'true quality', but judgement is all that we are left with. In the words of a senior scientist I knew, high-quality research made him 'tingle'. Another explained, 'the idea of excellence as a measure of research quality makes many people uncomfortable . . .[but] these people — despite their discomfort — cannot suggest anything better'.[8] So much about the judgement of quality depends on the integrity of the scientists undertaking research.[9]

In Pirsig's narrative, quality was the motor mechanic quietly going about his business fixing the motorcycle and turning this activity into an art form. Quality was something innate within the mechanic; he either had it or he did not Quality was not something which could be easily analysed, broken down into components and taught, like the components of the motorcycle and how they fitted together. But quality in the end is indicated by a whole, working motorcycle within which is embedded all the feeling and care permeating the machine from the actions of the mechanic. Pirsig proposed that the Platonic way of thinking about problems in the round was more powerful than the Aristotelian way, which focussed on instrumental analysis and functional forms. The Aristotelian way is closely associated with the traditional reductive style of scientific exploration, although in contemporary science the pendulum is swinging away from reductionism to embrace holistic concepts such as the emergent properties of systems, like an independently thinking 'beast' at the heart of the politics factory. Just knowing the components of something, like a motorbike, and how they relate to each other does not add up to the quality of the whole such as the sounds it makes when running, the exhilaration of the ride, its 'look'. Like the motorcycle, something like the brain is much more than the sum of its parts. Just sequencing the DNA of an organism does not tell us how the organism works or even how it looks. This split between reductionist and inductive, holistic thinking applies equally to how scientists make judgements about their own qualities. Some like to reduce their judgement to a few metrics such as the number of citations a published work has attracted from other scientists, whereas others like to say of quality, 'you'll know it when you see it'.

Peer review remains the principal way in which scientific quality is judged, and it applies a 'you'll know it when you see it' method of judgement, but as a result it is a blunt tool. It is the main method used to judge the quality of research before it is published in the scientific literature. However, about 90 per cent of published scientific papers are never cited and about half of scientific papers are never read by anybody other than the authors, peer reviewers and the journal editor.[10] These are branches on the tree of knowledge which go nowhere, possibly because in the end they are found to be wrong. I am certain that the bean-counters and efficiency gurus within the politics factory would say that, rather than allowing them to wither, they should not be allowed to grow in the first place. But getting things wrong is what science is about, and wastage therefore becomes an essential part of the search for quality. Science is an iterative, self-correcting process. Perhaps this great pile of wasted effort really shows us that the system works, and if it was to disappear through drives for greater efficiency, then the beast in the politics factory would be freer to choose the research it wants rather than the research it needs.

As it is, people still game the system. Perhaps as many as 16 per cent of scientists could have been involved in a practice known as 'citation hacking',[11] the unfair enhancement of their reputations based on the number of times their research is cited.[12] There is a lot of evidence to suggest that a very large proportion of the research produced by scientists has some sort of significant quality problem, and scientists themselves think there is a problem of integrity in their own profession.

About 4 per cent of images presented within the biomedical science literature are likely to be fraudulent in some way.[13] This is comparable to the level of fraud or error one sees across other sectors of society. For example, the UK's National Audit Office estimates that fraud and error account for between 3 and 6 per cent of national government expenditure.[14] The Association of Certified Fraud Examiners reckons that organizations lose about 5 per cent of their revenue to fraud each year,[15] and this level of financial fraud happens even in the presence of stringent accountancy controls. About half of all this fraud comes from people who are highly educated. Most often the perpetrators are male[16] and they are tenured within their organizations, suggesting that once they have established reputations, then committing fraud and getting away with it becomes more likely. Accountancy firms claim that financial audits cost, on average, 0.1 per cent of company turnover, but this saves about 10 per cent of turnover being lost to fraud, a benefit–cost ratio of 100:1. Translating this to scientific fraud is tricky, but if we assume that basic human nature is the same among scientists as in the population as a whole, then in situations where science is heavily audited (such as using peer review), we can probably assume fraud exists at similar levels to those found in financial transactions. It is just guesswork to estimate how much fraud exists in situations where audit is weak, but it is likely to be many times greater.

Financial and scientific fraud can be designed to do great harm[17] when they are deployed strategically to create disruption. There has to be a question about whether the rise of predatory scientific journals – which apply very low standards of peer review, and which mainly come from within unregulated jurisdictions – and *paper mills* could be part of a wider strategy of economic and social disruption, particularly of Western science. The political regimes in some nations would like to weaponize science, not by inventing increasingly sophisticated forms of kinetic destruction but, instead, by destroying trust in science as a method for sowing the seeds of wider societal distress. This is the end point of striving for *subjective* rather than *objective trust* in science. It calls for much greater safeguarding of the quality and integrity of science.

Does the judgement about quality in science really come down to something as ethereal as a tingling sensation? Possibly, but the pragmatist in me is uncomfortable with leaving it at that. Accountability on the part of individual researchers for the quality of what they produce needs to be central to any system of control, but this can only work if the mark of quality is clearly visible to those who use the outputs of science. Labels such as the *peer-reviewed scientific literature* which once provided a mark of quality have been corrupted by the machinations of the politics factory and have largely lost their credibility. We need an international standard by which the products of research, and perhaps also the people (the research scientists) who produce them, are judged.[18] In the following chapters, I suggest several strategies which can be used to overcome the trend towards the total subversion of science within the politics factory while still sustaining the input of science to solve many of the most pressing problems facing humanity.

Part 3

Taming the beast

At present, there are only two main ways for scientists to influence what happens in the politics factory. One is to use specially designed gateways involving committee structures and advisers who act as brokers. This has the advantage of maintaining ideological distance between the scientists themselves and politics, but the disadvantage is that the elites inside the politics factory control access to these gateways and have no obligation to either listen to scientific advice or justify how it has been used or why it has been ignored or never requested. The second is for scientists to enter the politics factory themselves – to wrestle with the 'beast' itself or, more likely, be subdued by it. An immense number do this, often out of sheer frustration because the formal gateways simply do not work, but they often enter in deceptive or subliminal ways, even to the extent of deceiving themselves about what they are doing, convincing themselves they are providing objective input when in reality they are pursuing an ideological position.

This is when the *scientifico-poltical predicament* goes unrecognized and its negative consequences flourish. Both scientists themselves and those on the receiving end of scientific advice can become confused by the mash of interests involved, some open and transparent, some minor and immaterial, but also some which are deceptive and potentially dangerous. When science becomes embroiled in this way, there is a real danger that trust – *objective trust* – in science is eroded as a source of knowledge, and, once lost, trust is very hard to regain. Science also loses its purpose, which is to dilate the iris in the lens of politics to focus on the big issues, and to apply inductive rather than purely reductive thinking when solving some of the biggest problems facing people and the planet.

This final section of the book proposes a few ways of thinking about how scientific insights might become more integrated within the politics factory to minimize these dilemmas and maximize the appreciation of objective trust. This includes packaging scientific concepts as broad, logically consistent frameworks (or paradigms) which can guide whole systems of decision-making, partly to avoid needless politicization. It works on the principle that if everybody can agree to march in the same general direction, then small deviations along the route do not end up with everybody losing focus on the final destination. Classifying political problems correctly, so that they can be chewed on, and potentially solved, by scientific approaches, is also a focus, but so is how we can start to view politics itself as a scientific process of investigation, a way of iteratively testing ideas and rejecting things which fail while accepting things which succeed. Then there is a need to ensure scientists themselves understand what kind of contribution they are making because different forms of science emerge from different kinds of motivations, some focussed on public good, but others focussed on private good. Often scientists themselves fail to understand how their contributions fit into this picture. Being explicit about these different motivations is important because this impinges on the judgements about trust in science. Finally, I look at the kind of governance mechanisms which could exist to make sure that science 'cannot just be dismissed on a political whim'.

13

Playing the paradigm game

In its established usage, a paradigm is an accepted model
or pattern. . . . Like an accepted judicial decision in the
common law . . . normal science possesses a built-in mechanism that ensures the relaxation of restrictions that bound
research whenever the paradigm from which they derive
ceases to function effectively.

Thomas Kuhn[1]

Thomas Kuhn made a major contribution to the philosophy
of science in the post-war era. He viewed the sociality of
science as a kind of tribalism involving groups of scientists
who came from certain intellectual traditions supporting
particular sets of scientific conventions. He suggested that
one system of thinking, or paradigm, eventually dominates
based on accumulated empirical evidence, but even this
eventually gets rejected as another, better paradigm arises.
It is in this way that knowledge rolls on and on to become
a progressively better description of reality. We no longer
think that the fish in the ocean are so numerous as to, in
effect, be inexhaustible by human exploitation, as was the
case in the late 19th century;[2] Newtonian mechanics was

replaced, at least for some applications, by Einstein's theory of general relativity; creationism and Lamarckism were replaced by the theory of evolution by natural selection; the James Webb Telescope is changing our perceptions of the early universe and how it formed; as we explore Mars, our paradigms for Martian lifeforms have changed, and so on. Kuhn's suggestion was that science *shifts* its paradigms often during periods of rapid evolution in ideas.

In the 1940s, Robert Merton had been at the forefront of helping to understand science as a social process involving a number of 'norms' of behaviour – *communism, universalism, disinterestedness, originality* and *scepticism.*[3] These turn out to be somewhat idealistic, and overlook *pragmatism*, but Kuhn's contribution built on Merton's ideas and was to link these kinds of sociological constructs and linguistics to how scientists actually worked. Scientists coming from traditions working within different paradigms will often disagree, but the ensuing scientific debate has been compared with a set of lively conversations backed up by observations and counter-observations of real-world phenomena.

There are parallels between this and what happens in politics, where arguments also move on, possibly based on the same kind of social process. Absolutism represented by autocratic monarchy was largely displaced by constitutional monarchy or republicanism underpinned by social democracy; as an experiment, national socialism was an unmitigated disaster; Marxism has largely been abandoned for other forms of social emancipation and various forms of capitalist economics from monetarism to Keynesianism; and, in the present, the rise of populism is seen as a significantly different political paradigm to that of liberal democracy. Current paradigms of liberal democracy involving consumerism, free trade and capitalism are coming under pressure from environmentalism. These can all be seen as experiments in social organization and governance which have failed, or will fail, in various ways, just as paradigms in science eventually fail. They are part of a rather sobering and perhaps fatalistic perspective in political science described by Hans Morgenthau and summarized by Henry Kissinger, as we saw in chapter 3, as 'the inevitability of tragedy'. The sequence of trials of different political solutions represents the painful evolution in political science, which has a lot of equivalence to the evolution of scientific ideas. This relationship was not lost on Kuhn, who called the changes *scientific revolutions*, a metaphor taken from the political revolutions of the 20th century. Even the natural world itself may have evolved in these ways, something known as *punctuated equilibrium*,[4] meaning that the presence of extended periods of stasis in things like species phenotypes tend to be interspersed by periods of rapid change, often driven by rapid shifts in environmental conditions. Overall, this is the consequence of something called non-linear dynamics and it is a strong, perhaps universal, 'what is true works' characteristic of how all complex political, social and natural systems evolve through time.

In Kuhn's view, scientists working within a scientific paradigm are involved in *normal science*; in the same way, politicians working within a political paradigm could be seen as involved in *normal politics*.[5] As Merton would have recognized, the social processes of science and politics appear to be very similar – two worlds governed by the same kinds of social interactions travelling in parallel through time.

It is important to emphasize that *normal science* in the sense used by Kuhn is not the same as *normative science* as described in chapter 10, which represents science done to confirm current preferences, but scientific paradigms themselves can be strongly normative. If followed through too strongly – especially when the scientific paradigm gets taken up as a political paradigm – they can give rise to *normative science*. An important difference, however, between politically led and scientifically led paradigms is that scientific paradigms have a presumption of fallibility – they will always be wrong in some way and will often be expected to fail. Scientists should be humble about their paradigms (although often they are not). In politics, individuals are rarely humble about their paradigms, and while they mostly know that their ideas are fallible, they behave dishonestly as if they are true, and the result is that they rarely plan for failure.

There are lots of examples of scientific paradigms which have become political paradigms. In fisheries management, it is essential to base everything on the quantity of the natural resources available because everything about the success of fisheries flows from the number of fish that are available to catch. This is determined using the scientific paradigm of stock assessment based on bio-economic models which are aimed at setting a level of fishing which maximizes yield over long time periods, something called *maximum sustainable yield*, also known simply as MSY. Invented in the 1960s, MSY reflects the productivity of Nature and has been built into fisheries policies ever since – the scientific paradigm has become the policy because to do otherwise, by determining the productivity of the sea based on human preference, would be plainly stupid.[6] MSY has, therefore, been built into the fisheries management policies of most nations.[7]

Similar paradigms exist for the management of infectious disease. The *epidemiological paradigm* aims to minimize the transmission of disease from infectious to susceptible individuals and is the basis upon which the World Health Organization operates as well as the equivalent international organizations for controlling plant and animal diseases.[8] Different tools have been developed to pursue the *epidemiological paradigm*, from vaccination of susceptible individuals to killing infectious individuals (e.g. the culling of badgers as described in chapter 3) and the restriction of contacts as used in 'lockdowns' or 'social distancing', which happened during the COVID-19 pandemic.

The *gross domestic product* paradigm – shortened to GDP – drives the economic objectives of many nations and many nations also organize themselves around a social paradigm of *representative government*. A universal paradigm involves *money as a token of exchange*, a paradigm which is being strained more than at any time in the past as a result of the decoupling of money from tangible assets and the rise in cryptocurrencies and cryptoassets.[9]

Some paradigms can be deeply socially and structurally embedded, such as the paradigm involving how food is produced by farming. In this case, it is so powerful that it holds back scientific innovation to shift the paradigm, resulting in an increasing proportion of the environmental impacts of humanity coming from food production. Slavery was once an economic paradigm which took about a century to shift, although, even if it is no longer mainstream, it still lingers on, cloaked in supply chain mist.

We also have paradigms which are in the ascendant, like the paradigm of *human-induced climate change* and *global biodiversity loss*, or the paradigm of *planetary boundaries*, which holds that there are limits to the use that can be made of the services provided by the planet to support the demands made of them by people. These are competing with old, increasingly outdated paradigms involving, respectively, a *stable climate* and *unlimited growth of humanity*. Shifting scientific paradigms like the shift from analogue to digital technology can also shift society, in that case because of its impact on how we communicate and use data and information. Much the same can be said for the effects of shifting paradigms in reproductive endocrinology which led to pharmaceutical contraception and the emancipation of women, and for paradigm shifts in vaccine technology in the form of new malaria vaccines which are on the brink of changing society within sub-Saharan Africa.

The purpose of science is not specifically to shift paradigms but is to confirm that current paradigms are still fit for purpose and to continually test whether alternatives are possible. Science creates options for which paradigm to follow. However, we can become so heavily invested in some paradigms that they become difficult to shift, even in the face of overwhelming scientific evidence that they are no longer useful. Like slavery, they hang around like acid reflux from a bad meal. As early as 1977, there were those who thought MSY in fisheries was unhelpful,[10] but it has such a strong presence in international agreements about the management of fisheries that it is now impossible to shift and is probably a significant cause of overfishing. Some paradigms can also verge on ideology, such as Keynesian economics or monetarism. When those paradigms start to be associated with specific names, such as Milton Friedman with monetarism, then they take on the complexion of personality cults and tribalism ensues. They lose their objectivity and people start to behave as if the paradigms are absolute truths. Objective trust in the paradigms is transmogrified into subjective trust.

Many of the paradigms we use to manage and govern us have failed and are now no more than a cultural or administrative millstone. Trickle-down economics is now largely discredited but still forms a backbone of the economic policies of nations and attracts surprisingly wide support even among people who should probably know better. Outdated paradigms get baked into policies and cultures in ways which make them hard to get rid of. Getting rid of MSY would be desirable but would also be highly inconvenient because many nations have built industries around getting access to certain amounts of fish using it as a basis for their calculation. It thus becomes difficult or impossible to un-bake the poisoned cake made from the ingredients of an old and outdated paradigm. MSY has become in effect a flawed truth – like Santa Claus – where the flaws are conveniently glossed over. A consequence is that a lot of fisheries science descends the subjective by-pass to become normative. Rather than seeking novel, disruptive and better solutions, many fisheries scientists spend their whole careers ploughing the same old rut, wasting their talent and promoting decisions which often lead to over-exploitation.[11]

Another area of research where I have seen this happening is in agriculture and food, where there is a desperate need for out-of-the-box thinking to solve the global strategic challenge of how to feed the global population without also ruining the environment. Instead, talent and money are wasted and opportunities are lost because scientists become channelled into addressing questions which only fit with current paradigms about how we produce food for people.

Many international agreements, such as the Convention on Biological Diversity, the Convention on Climate Change and political conventions like the UN Sustainable Development Goals, tend to bake in the paradigms which existed when they were created. These will often be found to be an unhelpful and inaccurate view of the world as experience and science move on. But changing them once they have been accepted is notoriously difficult. For example, most of them were agreed before we started to conceptualize Nature as a capital resource – a resource whose products are needed by us for survival[12] and a resource which we, therefore, need to maintain rather than progressively degrade.

I explained in chapter 3 that badger culling was a mis-specified problem. There, the *epidemiological paradigm* had been rolled out to control bovine TB, often supported strongly by scientists, but this has increasingly looked dated. Instead of looking at the reasons for badger culling as an epidemiological problem, it looks much more like it is a human social problem requiring very different solutions from shooting badgers, such as changing the structure of incentives for farmers.

Neither badgers nor fish read, understand or respond to policy documents, but people do, and it is the social rather than the natural processes which can be managed, whether this concerns fish, badgers, mineral reserves or viruses. The mis-specification of the scientific paradigm underlying policies often occurs because there is a wish to avoid confronting the fact that most problems in policy concern people and solving those problems usually concerns changing the behaviour and expectations of people rather than trying to reshape Nature to people's wishes. The idea that we actually are able to manage fish populations, which is an underlying assumption of the MSY paradigm, has proven to be mostly untrue. What we do, in reality, is manage the people who fish, just as we manage farmers to control TB or people to control COVID-19. A major strategic paradigm shift over the last twenty-five years has been to stop imagining that we have the power to manage Nature and to start imagining that solutions lie in managing the activities and expectations of people.

This conceptual shift creates the circumstances for the constructive merger of science and politics. It brings science to the front and centre of the debate about how to live. Paradigms provide the materials to construct narratives with which people can engage, where a 'narrative' is a story or expression which simplifies and contextualizes a very complicated flow of information so that people can make sense of it. Metaphors are a part of this picture, as famously illustrated during the COVID-19 pandemic by Anthony Fauci, the Director of the National Institute of Allergy and Infectious Diseases, in the US and Jonathan Van-Tam, the Deputy Chief Medical Officer, in the UK, who used metaphors to communicate complicated systems of ideas, or paradigms, through familiar parallel illustrations.

The basis of paradigms (even for scientists, who, like all humans, have limited abilities to hold all the information they need in order to fully appreciate the intricacies of many of the systems they study) is that they are narratives which help to create general alignment between different actors, many of whom struggle to achieve a common understanding using pure linguistic descriptions. I learned when working in the politics factory that there was no limit to the number of times that even quite simple ideas had to be repeated in order to ensure that a critical mass of people actually understood what was being said. By reformulating messages in multiple ways, they are more likely to appeal in many cognitive circumstances. People latch on to metaphors much more readily than to abstract or uncontextualized concepts because our brains appear to be constructed so that we appreciate very complex processes better through stories. Indeed, as Yuval Noah Harari has pointed out, story-telling is perhaps our most powerful trait.[13]

Creating narratives from paradigms can be a useful way of transferring what goes on inside science to create resonance within politics. Some critics even felt that Kuhn made science itself sound like a political affair,[14] one where socially accepted scientific leaders were resorting to rhetoric to defend specific scientific positions; the fields of climate science and GMOs have been troubled in this way. But narratives can help to open minds without hectoring. 'One-third for the birds' was a strap-line slogan used by me and colleagues when talking about the research we did to define what we thought were more defensible exploitation rates for the ocean (see chapter 3).[15] We were proposing an entirely different paradigm from that of MSY to set the level of exploitation. A complicated train of logic was explained in just five words, something which was more likely to have resonance within the politics factory than the mathematics, graphs and thousands of words of carefully drafted scientific argument which sat beneath this message.

In contrast to scientists, politicians are very good at generating narratives which people find attractive, and even if they are often dishonest, it is what makes people vote for them. Scientists also need to learn to generate the narratives which are attractive to people but which, as faithfully as possible, reflect the scientific paradigm, something they have traditionally struggled to do. But progress is being made. The UK Climate Change Committee presents scenarios which would allow transition to a low-carbon lifestyle and which are, essentially, narratives about what needs to change.

Scenarios are alternative illustrative narratives about how the future might unfold and are a method used to communicate scientific messages within the politics factory. The IPCC has projected future scenarios for climate response to changing levels of CO_2 in the atmosphere depending on what policies are implemented. Models were used during the COVID-19 pandemic to develop scenarios depending on which policies were implemented to control viral spread. Sometimes these projections are expressed as a *reasonable worst case*, which often reflects the way the future will evolve if nothing is done, or if we are down on our luck. Scenarios have the capacity to engage people with complicated concepts and allow them to understand where trade-offs might exist in future and what these mean for them. This simplification can be an effective way of helping people to understand how scientific paradigms can beget new political paradigms.

As a scientist, it has always been my wish to help lead society-wide conversations about how we should live based on the best available knowledge, as well as its uncertainties. I thought that my job, as an adviser to government ministers, was to cut out petty politics and to help create the headroom often sought by broad-minded politicians to bring about structural changes across society. The main mechanism for this has been to provide objectively argued scenarios about how the future might unfold under different sets of assumptions. A major barrier to achieving these conversations has always been the reluctance of the politicians I have advised to release those scenarios to the public because of the profound challenges they would present to people's lives and their conceptions of themselves.

The argument so far has suggested that scientific paradigms beget the narratives which build policies, but there are circumstances when stories might beget scientific paradigms. Storytelling by indigenous peoples which derives from traditional knowledge systems might indicate valid scientific paradigms.[16] There are good examples of science gaining new insights from listening to these stories, but it is always important to remember that whereas parts of these stories will stem from real-world knowledge, other parts will be useful fictions, like Santa Claus, and it is often not possible to know the difference. I recall one instance of Inuit hunters of Beluga whales telling stories about 'their' whales turning up year after year at the same location and that they therefore had ownership rights to hunt those whales. Subsequent scientific studies showed the idea that these people had exclusive ownership of the whales was fiction because the same population of whales was being exploited by several different Inuit groups across the Arctic and all of them thought of them as 'their' whales.

Storylistening, as it is sometimes called,[17] is certainly an important part of the skills set for scientists, but rational scepticism needs to be applied to any story because paradigms presented by people may be just as likely to be constructed to deceive as to inform.

Some political scientists of the 20th century like Charles Merriman and Quincy Wright proposed the instrumentalization of policy through the pursuit of paradigms. But when taken to the extreme, absurdity can emerge, like the 'body count' objective set out by US Secretary of Defence John McNamara during the Vietnam War. Just as ridiculous, although not recognized as such, is the obsession with GDP as a measure of economic objectives. These examples follow the mantra of that great 19th-century scientist Lord Kelvin, whose view, as we saw in chapter 4, was that if you can measure something, then you begin to know something about it. Although certainly true, in the context of systems science, these measurements should involve multiple factors reflecting different components of the system. Mathematical methods can be used to combine them into broad indices of progress towards system-wide objectives. So, during the COVID-19 pandemic, one could have easily zeroed in on measures of disease prevalence in populations, or the number of new cases of disease, as measures of success. They were certainly important, but they needed to sit alongside measures such as the employment rate, hospitalizations and business performance. This is because policy implementation is a process of optimizing among many different factors which are often in a trade-off with one another.

Paradigms, policies and scenarios may seem different concepts, but they all carry a common strand, which is that they are about telling stories and turning these into lived experience. The paradigm is the scientific story; the policy is the story of how collective action, often under the supervision of government, might apply scientific knowledge; and the scenario is the story about how different policy options are likely to play out. When used in combination, these can be effective at building science into the activities of the politics factory. Playing the paradigm game is a key skill for scientists to deliver societal relevance and impact.

14

Taming wickedness

'Wicked' problems can't be solved, but they can be tamed.

John C. Camillus[1]

Unlike academia or commerce, governments don't generally get to pick the problems they have to solve. Seen from the inside, problems seem to come at governments with relentless frequency and vigour and there is never enough time to think about how to get ahead of them or to stop them materializing. Politicians try to put a good face on this, but when one is inside government it often feels like being on a stricken ship. As a scientist searching for solutions, I thought this was partly caused by an under-appreciation of how strategic thinking and systematic methods of observing the structure and typology of problems themselves can help to bring about solutions.

One way of thinking about problems is that they have two divergent components involving, respectively, their immediate relevance and their deeper implications. Even if a political or policy problem has proximate relevance, it almost certainly illustrates something fundamental in terms of its root causation. Sometimes these contrasting elements are known as the *applied* and *basic* components. Both these components are almost certainly present in every problem, even if the motivation for searching them out can vary greatly. Scientists can place particular value on investigating and understanding the *basic* components, but the *applied* components are often most valued by people who simply want a solution to their immediate concerns. In general, this is the attitude taken by people within the politics factory.[2] Often scientists themselves are employed to deal mainly with *applied* problems, but if they lose sight of the *basic* components, then, in a sense, they sell their soul and are in great danger of becoming a simple instrument of politics. In contrast, scientists who seek only *basic* components are in danger of losing their social licence to operate.

Illuminating generality from working on specific problems is one of the hallmarks of great scientists. Science investigating the *basic* component of problems feeds *strategic* solutions (which are broad and long-term), whereas *applied* science feeds *tactical* solutions (dealing with the here and now). A mistake often made within the politics factory is to seek only *tactical* solutions and not to look deeper into problems to seek out fundamental truths or lessons. Badger culling, for example, was a tactical solution which resulted in most people involved ignoring some of the fundamental truths about how bovine TB was transmitted. Apart from the fact that these superficial approaches to problem-solving rarely come up with robust and lasting solutions, the lessons learned are rarely transferrable to homologous circumstances where the same root causes are in play.

In my view, this political disinterest in the fundamental components of problems is largely why being inside government feels so much like being on a stricken ship. But also in my view, it is the job of scientists to ensure that the stereotypical focus on *tactical* solutions from within the politics factory is exploited in order to deliver co-benefits in the longer term by also delivering *strategic* solutions built on *basic* knowledge. For example, it is one thing to measure the size of a fish stock to set the level of fishing in the immediate future (the tactical component), but exactly the same measurements can be used to research fish stock dynamics in general (the strategic component), leading to better policy formulation to establish exploitation levels which address long-term conservation and economic objectives.

Similarly, a common complaint is that medical intervention focusses too much on the treatment of symptoms when a much more generic approach to prevention of root causes of disease would bring long-term, cheaper and fairer outcomes. The politics factory has a strong tendency to want to treat symptoms of problems rather than underlying causation, and I suggest it is the job of science to champion a greater focus on the root causes.

However, shifting the focus from tactical to strategic solutions can be difficult. As one national leader said to me when I was advising about setting objectives to tackle sustainable consumption, 'There is a clear imperative to achieve these objectives, but they will only be achieved if we take people with us.' She was recognizing that even if strategic solutions were essential, people were mainly responsive to tactical interventions. The political imperative is the 'what's in it for me?' question posed by voters, and scientists need to be just as prepared to formulate solutions to this question as they are to forage for root causes and to champion them.

Another angle where science can promote solutions to problems is to coach people within the politics factory about what kind of problem is being tackled and where it lies on the spectrum of tractability. Knowing the kind of problem being tackled can lead to the adoption of very different methods to manage it. The function of science should be to march problems from intractability to tractability, from insolvability to solvability, and this process starts by classifying the problem in terms of its tractability. Knowing what kind of problem one is handling can make a big difference to how well science can intervene to improve the situation, and can even sometimes determine that science cannot help at all, especially when problems have become *wicked*.

I found that progress in this respect can be made by applying a typology of problems along the *tame–complex–wicked* spectrum.[3] I also doubt that science ever makes *discoveries* in the sense that new things simply pop up out of the blue. Instead, researchers try to solve problems and in the process new observations emerge. These can seem completely novel and unexpected to the uninitiated, but they emerge from a rational process. This problem-solving focus of science is really no different to what happens in fields of business and public administration where politics plays a central role.

Problems which are already *tame* are inherently solvable, and much of science has developed to deal mainly with tame problems. This does not mean they are easy to solve, but they have a defined solution which, with focussed, structured actions, means they *can* be solved. Defining how proteins fold when they are created is a tame problem because, even if it is very difficult to understand how protein folding happens and to predict the folding behaviour of proteins from first principles, it has a single definable solution. Getting a man on the Moon was a tame problem. By throwing enough resources and brainpower at the problem, it could be solved. There is no point in defining equivalent 'moonshots' to solve problems which are not tame. For example, curing cancer is not a tame problem because cancer is a catch-all term for a wide range of different genetically derived diseases which affect cell division.

So, declaring a cancer 'moonshot', as President Biden did in 2022, is really rhetorical nonsense.

Some of the foundational problems which have been solved in the modern age are tame in nature. There are well-known examples like solving the problem of how inheritance works through information encoded within the structure of DNA, and how matter is structured by investigating it using big machines like the Large Hadron Collider. The observation of gravitational waves has taken us a step closer to solving the problem of what holds the universe together. The development of antibiotics, vaccines, lasers, hormonal contraceptives or semiconductor materials has formed the basic tools to allow us to solve problems from fighting infectious disease to improving the speed of communications and achieving social emancipation. As I described in chapter 1, progressive refinements of the problem of how to measure time has shaped, for better or worse, our societies and politics. There are many more examples, but these are all problems with a final solution, a clear end point.

Tame types of problems which promise future revolutions and which have yet to be solved include some in the field of synthetic biology, quantum engineering, artificial intelligence and nuclear fusion.[4] Future access to sufficient energy for the planet without use of fossil fuels is a tame problem, even if it has yet to be solved. Putting a person on Mars is also a tame problem. Tame problems can be solved by traditional scientific methods including experimentation and hypothesis testing. However, not all tame problems are solvable in practice. It is possible that putting a person on Mars becomes impractical because human physiology and psychology cannot cope with the rigours of travel and isolation. I recall once considering a tame problem which I knew how to solve, but it would have taken much more than all the computing power on the planet to find the solution.

Sometimes one has to give up and admit defeat even with tame problems.

Complex problems are very different in that they often have no precise solution, but they are very important for governments to understand and work on because they are probably the most common kind of problem encountered by civil servants and politicians.[5] They are the kind of problem that you could throw endless computer resources at without really being sure what the outcome will be, even if we know all the underlying causation. The fact that all zebras have unique stripes is the emergent property of the developmental process which happens in the ectoderm of the developing zebra embryo. Even if we know all the components that make up a zebra, we cannot precisely predict the pattern of its stripes. Very small changes in the arrangements of molecules in the primordial cells which eventually develop into the pigmentation patterns which create the zebra's stripes result in different outcomes. Our own features – our phenotype, or how we look – are the result the same kind of *complex* process and it is why we all look different.

Complex used in the sense meant here is not the same as the common meaning of the word, which refers to something which is bewildering. Instead, it refers to a type of problem which evolves through time and where there are often many interacting components. If we consider all those components as a *system*, like all the cells in our body represent a system which we see as a living human being, the state of the system in the present is correlated with its state at some point in the past, and the same can also be said for its state into the future relative to its state now. Random systems, like the games in which dice are thrown to dictate outcomes or the radioactive decay of an atomic nucleus, have no memory, but complex systems do have memory. In fact, in Nature, there are very few truly random systems.

Most are complex in some ways, and many are much more complex, and therefore more difficult to predict, than we imagine.

In physics, something called the n-body problem, which is used to predict the motion of celestial objects, shows us that when a system contains two interacting bodies (when $n = 2$), then its behaviour can be predicted exactly, but when more are added, the calculations become too difficult to solve exactly. When n becomes very large, as it does for the whole of the solar system, including all the planets, asteroids and comets, it can appear chaotic when viewed over long periods of time. This is why meteorites from Mars sometimes end up on Earth and why an asteroid ended up hitting the Earth about 64 million years ago, which caused the extinction of most of the dinosaurs.

But if we replace celestial bodies with human bodies, the n-body problem is one reason why social interactions are also so difficult to predict. Human society is a complex system, as is government. Even if it is rarely explicitly acknowledged as such, government attempts to manage complex systems, often without adequate tools for the job and even a lack of appreciation that what is being managed *is* a complex system.

One of the most common misjudgements which happens within governments concerns the misclassification of complex problems as tame problems, something which stimulates the belief that investments in science can lead to a definable solution. This is a road which just leads to disappointment and disillusionment with science, but it is a symptom of problem mis-specification.

We have a certain incapacity to predict futures in complex systems because of the need to know the arrangement of the components of a system to an impossibly demanding level of detail. The mathematical meteorologist Edward Lorenz famously pointed out that the flapping of a butterfly's wing is enough to determine the ultimate size, strength and path of tornadoes and hurricanes.[6] The variation of water flow in rivers, the dynamics of ecosystems and the organization of growth in plant and animal populations are all typically complex processes which we struggle to predict.

In the sociosphere, the flow of money in the banking system and the flow of information through social media are also similarly difficult to predict. Some complex systems sit across the boundary between the sociosphere and the biosphere, such as the global food system, which relies on natural processes to produce food as well as markets to trade that food, and their progression is impossible to predict even over short time-scales. The observed state of all these systems in the present or the recent past provides some information about what they will do in the future, but as we stretch our projections of their behaviour out into the future, our capacity for prediction diminishes until it reaches a point where it is no better than a random choice. This is the *prediction horizon* for a system, also known as the *Lyapunov time*. We put huge amounts of effort into predicting the weather, which includes some of the best research in fluid dynamics and meteorological physics, all brought together in simulation models run on supercomputers, but none of this can ever provide a weather forecast with any accuracy beyond about fifteen days into the future. This is the *prediction horizon* for weather forecasting and, as far as we know, we can never improve on this.

When the British politician Harold Wilson supposedly said that 'a week is a long time in politics',[7] perhaps he was reflecting on the Lyapunov time in the socio-political system. The relevance of all this to politics comes from the fact that governments spend most of their time managing complex systems, and political leaders are held accountable for the management of those systems on behalf of the population. These are manifest as systems for managing health, finance, transport, food, education, defence and many more. When these are all brought together alongside the social systems used to debate how best to run all these systems, often described as politics, then one has a system of systems. Deterministically managing these systems, as if they are composed of solvable problems, as governments are prone to do and politicians often seek to do, and their voters usually expect them to do, is an utterly foolish exercise.

Too often in politics, *complex* problems are misclassified as *tame* problems, creating expectations of solutions which are simply unattainable. For example, in 2020, the press wanted to find all sorts of simple explanations for differences in the progression of the COVID-19 pandemic between nations. However, because this was a complex process, even very small differences in the starting conditions of the pandemic in each country could have accounted for the radically different trajectories of the pandemic in each case. This insight alone should really have killed off political and journalistic hyperbole about which nation's policies for managing a pandemic were the best.

Governments could be a lot better at managing the systems they are responsible for if they were to use the understanding we have of complex dynamics. This might bring only a very modest improvement in the capacity to forecast the future because inherent instabilities mean the *prediction horizon* is never far ahead, but it might make people more realistic about the uncertainties which do exist and encourage them to plan accordingly. Techniques involving 'nowcasting' are increasingly helping to manage systemic processes. These use tools like artificial intelligence to make forecasts across a wider range of relevant activities from food prices to the flows of people in transport systems or the use of hospitals.

Even if complex systems might seem unfathomable, they do have structure, and just tracking this is what makes them tameable. These systems often oscillate around things called *strange attractors*. The planets do not orbit the Sun, as commonly taught, but instead they orbit around a common point, their attractor, which is the centre of the gravitational pull of the solar system. These attractors can be envisaged by comparing the relationship of components within a complex system to people on a dance floor. It is hard to predict at any particular time where any individual will be on the dance floor because it looks like they are moving around chaotically, but all those on the dance floor in fact are moving around a central point, the *attractor*, probably placed quite close to the physical centre of the dance floor. If this central point moved (i.e. if we moved the dance floor), then the dancers would move with it. The dance is like a complex problem, but we can see organiza- tion within it. Similarly, life, defined as a living organism, is the emergent property from a complex process moving around its *strange attractor*. Even if they are somewhat abstract in their conception, these *strange attractors* are incredibly abundant in Nature and they are the glue which keeps it and us from falling apart.

The dynamics of what is going on inside government have the same properties, and just being able to step back and watch the dance with these insights marks out the dif- ference between great leaders and people who get tied up in details. As Napoleon (perhaps the greatest ever strate- gist) said, 'strategy is the art of making use of time and space', which is like choreographing the dance, in his case, of armies around Europe. Napoleon undoubtedly had sig- nificant flaws but, despite his eventual downfall, he knew instinctively how to manage systems.

When systems like a government tend to oscillate around a consistent attractor, then they can be described as being

in *equilibrium*. We like equilibrium systems because, even
if they can be hard to predict in the short term, in the long
term their state is broadly predictable. Our food system has
a lot of the characteristics of an equilibrium system. Trust
and reciprocity in financial, social and political networks
are essential to maintain the order and structure of human
society but, as we know, these can break down quickly,
involving the potential loss of life-critical processes like
food distribution.[8] It would be a disaster if our food system
started to show non-equilibrium behaviour. Such a shift
happened to the global banking system in 2008. This was
an example of a system failure, commonly also known as
a *tipping point*. This is when the *strange attractor* moves,
often in surprising and completely unpredictable ways. A
tipping point in the attractor which holds together a living
organism is death. Politicians and policy jockeys who advo-
cate for rapid transitions, often referred to as 'reforms',
'resolutions' or 'radical changes' – sometimes 'revolutions'
– are instinctively trying to shift the position of the *strange
attractor*. This happened when the UK left the European
Union. My own home government in Scotland wants independence from the rest of the UK. But what these people
need to understand is that shifting the attractor of a complex system can result in highly unpredictable outcomes.

However, attractors within social systems can be so
strong that no policy is likely to shift the system into a new
state, even if governments want this to happen. Evidence
for lack of appreciation of this is all around us. Efforts to
reform healthcare systems are notorious for their failures.
The politics factory manufactures ambitions to shift these
systems while failing to account for the immensely influential stabilizing feedback which exists across society and
which stops them from changing. For example, ever since
the European Union was created, there has been an ambi-
tion to reform agriculture. Superficially, agriculture is about

producing food, but when translated into the dimensions of politics, it is a heady mixture of health, economic, social and environmental policy with huge social capital invested in each by major vocal and fanatical factions across society. All the various pressures which exist within this policy domain push and pull at each other, making it impossible to shift the beast, or the attractor, so nothing changes. Reform is impossible in these circumstances, at least when using the normal tools available within the politics factory. Only a major kick by some sort of catastrophic event or by the development of some form of revolutionary new technology is likely to lead to change.

These kinds of insights tell us that people within the politics factory often ride a wave built on unrealistic optimism about their capacity to bring about change. When this happens, however, there is little control over where the attractor might settle and there will be a sense of impending chaos while it drifts to a new stable place. Simply recognizing the existence of system complexity might instil a better sense of proportionality within the politics factory about what is possible and about how change can be managed. When this call for proportionality and thoughtfulness about the nature of problems fails, then *complex* problems turn into *wicked* problems, and turning complexity into wickedness is the greatest and most frequent failure of the politics factory.

Wicked problems are a very different challenge from either *tame* or *complex* problems and, sometimes, it is best for scientists just to step away from them. Often they are beyond help. One of the earliest references to wicked problems suggests they refer to 'that class of social system problems which are ill-formulated, where the information is confusing, where there are many clients and decision-makers with conflicting values, and where the ramifications of the whole system are thoroughly confusing'.[9] Not only are they problems with no solution, but they also intentionally defy efforts to find solutions. If Nature was truly chaotic, in the semantic rather than the mathematical sense, then it would present us with wicked problems. Fortunately, Nature has structure which is unlikely to fall apart at the least provocation. I am not sure if *wickedness* is a purely human trait, but it is certainly very human-centric.

Wicked problems might be characterized as chaotic problems with no attractors or where the attractor moves with little provocation – where the dance floor and the dancers are both moving at the same time.

The mind boggles when trying to imagine what Nature would look like if it presented wicked problems. For example, even if the average flow of a river was unchanged, it might one day be overflowing its banks and the next day be an almost dry runnel, and, to make matters worse, it would change its pattern of flow to wilfully circumvent any constructed flood controls. It would display behaviour which was not just erratic but intentionally obtuse. Its flow might even be unrelated to rainfall.

Perhaps the closest we get to understanding this in daily life is the way in which cancers, once exposed to chemotherapy designed to kill them, evolve to resist those therapies.[10] In doing so, the cancer loses sensitivity to any therapy and grows uncontrollably, killing the person it relies upon to sustain it. We certainly see cancer as wicked, but the science undertaken to find cures is attempting to tame that wickedness. I do not think it is too dramatic to compare a lot of what goes on in the field of politics as having equivalence to the dynamics of cancers. People have a capacity to indulge in self-destructive activities, something sometimes known as the *idiot index*.[11] An example might be striking workers, who, by striking, create the conditions where the jobs they want to protect become less, rather than more, secure. There is no hope of justice when wicked problems flourish. They reflect the mendacious side of the 'beast' at the heart of the politics factory.

One of the strongest features of wicked problems in my experience is that the act of studying them becomes part of the problem itself. They are often created by unresolvable trade-offs, such as where there is recognition that large wealth gaps lead to conflicts, but there is little appetite to address wealth disparities and the existence of these disparities may even be the reason for rising wealth in the first place.

A golden rule for any scientist, I suggest, should be never to make a wicked problem worse. This is an extension of *primum non nocere* – 'first, do no harm' – which is a concept of moral principle most often associated with medical practice. Often, at a personal level, this will translate to a practical assessment of whether a problem is *wicked, complex* or *tame*.[12] When scientists become a part of the problem they are studying, often by conducting or promoting *normative research*, then they are much diminished as channels through which problems can be tamed. A scientist in the politics factory needs to have the prescience, capacity and detachment to understand how to recognize and classify problems and then to build a balanced set of both tactical and strategic solutions.

15

Adaptive policy testing: making policy into science

Adapt or perish, now as ever, is nature's inexorable imperative.

H.G. Wells[1]

Earlier in this book, I suggested that one of my objectives is to demonstrate how it might be possible to make politics more like science. As I have shown, doing the opposite by making science more like politics is a certain pathway to disaster. I now what to explain what I mean.

Both *complex* and *wicked* problems defy the application of reductive logic to find solutions. Indeed, solutions in any normal sense do not exist for these kinds of problems. Since government is full of these kinds of problems, the standard linear methods involving policy creation and execution are inadequate to make progress. Instead, there is a need to build knowledge about problems and to manage them in ways which help to converge towards a solution of some sort, even if all concerned accept that specific, singular solutions are idealistic.

This is a process which could be called *adaptive management, adaptive experimentation* or *adaptive policy testing.*

If a policy can be equated with an idea or hypothesis and its implementation with an experiment to observe what effect it has, then we are a long way towards defining politics as science. The single important additional ingredient needed is for the outcome to be compared with expectation and then, having sought rational explanations for the inevitable departures from expectation, to reformulate the policy and run the experiment again. It is this step which politicians find so difficult because many of them stake their reputations on the experiment working first time. This is part of the stupidity of politics: the making of promises which cannot be kept. However, greater involvement of science in the political process would help to alleviate this, even if it might fail to stop it happening altogether.

The *adaptive* method applies rational scepticism to outcomes and it aligns with Peirce's tough-minded 'what is true works' pragmatism. If 'experiment is the sole judge of scientific "truth"', as Richard Feynman said,[2] then this is a method of designing policies in ways which converge on truth. The *adaptive* method turns science from being the supplier of technical solutions into the creator and implementer of policy. This simple insight alone can shift how investments are made in science. For example, regarding the emotive issue of solving the problem of plastics pollution, the important scientific questions shift from a search for toxic effects to policy experimentation.

Adaptive policy testing is also closely linked to something known as *the policy cycle*, which describes the theoretical steps taken when creating, implementing, evaluating and revising a policy. But it differs from the policy cycle by being more than a heuristic description of what sometimes happens.[3] It is, or should be, a disciplined method applied with rigour, oversight and accountability.

Adaptive policy testing is not a new concept by any means, and most governments in advanced economies have this capability[4] but they have rarely deployed it effectively. I encountered people in government who were specialist policy evaluators who were simply ignored most of the time. I once made a valiant (but eventually fruitless) attempt to have policy success included in the government's audit procedures. Politicians dislike having their ideologically based policies assessed for effectiveness, and civil servants are also incentivized to big up success rather than to celebrate failure when it is the failures which provide important lessons. The Schumpeterian dynamic within the policy factory is merciless towards failure, even though failure is something which should be expected for all policies, just like it is for all scientific hypotheses.

Force majeure can bring the politics factory to its senses. Consider, for example, the COVID-19 pandemic, where it was adaptive policy testing which dug us out of the mire. Governments around the world had no option but to adopt a tough-minded stance when dealing with the radical uncertainties associated with the spread and impact of the virus. Different interventions were formulated based on the best knowledge at the time, tried out and then either strengthened or abandoned. Multiple interventions were introduced simultaneously: for example, social distancing alongside the use of face masks alongside border controls over movements of people. It is impossible to break these down to understand the effects of just one component. Whole systems of intervention need to be considered, creating challenges for making judgements about which components of a system (e.g. face masks versus social distancing) are being most effective.

One of the most important examples of *adaptive policy testing* I was responsible for involved the methods being applied to improve air quality in cities, where there is increasing evidence that poor air quality underlies significant morbidity among people. There was enormous uncertainty about how much different interventions (such as car scrap-page schemes or congestion charging) would contribute to achieving the aim of improving air quality. The whole issue had got bogged down in a political quagmire (see chapters 3 and 11), resulting in government paralysis, but, fortunately, the UK Supreme Court had insisted that the government had to act, because it would presumably be immoral not to. The result was the creation of a large number of low emission zones within many British cities, each one of which can be seen as a replicate within a large-scale and long-term experiment. The continual evaluation of the results against some simple measurements of air quality will eventually allow those controls which are effective to stand out, allow-ing fine-tuning of interventions towards achieving the medically recommended objectives.

The situation with air quality was ethically based. Even if politicians found it difficult, it would have been uncon-scionable not to act. *Adaptive policy testing* can help to resolve ethical dilemmas, thus reducing the chances of these problems descending into wickedness. The ethical problem for those involved, in very generalized form, is that there is a choice between two courses of action: either to act and to reduce the impact of something like a rampant disease or poor air quality; or not to act and allow these things to take their natural course, which would very likely result in many more people dying.[5] This involves each person having to flip a personal switch between the two different courses of action: to intervene or not to intervene.

When tested, most people think the right ethical action is to intervene to kill fewer people.[6] This is an action cap-tured in particular by virtue ethics, because instinctively the choice people feel most comfortable with is to take care, to be precautionary. They put themselves in the place of the potential victims and do those things which are going to lead to the fewest regrets. They want to save lives. Even if it was not made explicit, this was the ethical framework for most decision-making during the COVID-19 pandemic, at least during the early stage of the response, but things changed as the pandemic grew, when a more utilitarian approach emerged.

Utilitarian ethics has an especially strong presence in Western-style governments,[7] and as more information became available, people began to calculate (both formally and informally) the relative utility of different approaches. They began to consider things like the harm done to businesses and social relationships. Gradually, it became more acceptable to kill people because there were trade-offs between the costs and benefits of different actions. In the case of air quality, the judgment of the UK Supreme Court put utilitarianism to one side and it opened the possibility of building an empirical approach to solving that particular problem.

A third ethical framework which drove some other people's opinions during the COVID-19 pandemic, including those of many scientists, involved minimizing the interventions. In the vein explored within Kantian ethics, this can be the right moral action because choosing to intervene is a conscious action and whatever we do people will die. If we cannot be sure of the consequences, and especially if there are much larger uncertainties associated with intervention than we are sometimes prepared to admit, then the moral position is to be very cautious about what interventions are applied. This draws on the idea that the very act of acting makes you immoral, and anybody experienced with policy implementation will have experienced the law of unintended consequences. Over-confidence with actions to intervene is likely to result in unintended outcomes.

Kant's view helps us to make sure that when *adaptive policy testing* is applied, then it is done with all seriousness and any actions are built on the best available knowledge or analysis. Frivolous actions would simply be immoral. It eschews the idea that economic assessments commonly used by governments can describe the consequences of specific actions within reasonable bounds – as I discovered for air quality. It also stops us throwing resources aimlessly at the problem just because we feel duty-bound to do everything in our power to stop bad things happening.

Adaptive policy testing is a powerful method for taming problems, and especially for dealing with those which have high uncertainty, where there is political demand for results and where strong values-based judgements are in play. It applies the scientific method to politics; it is the application of Kant's view that 'knowledge begins with experience'.[8] Even if it is rarely acknowledged explicitly, much of government works in a mode which means it is not much different from experimental science. Ideas are being tried out and tested, then modified based on experience, and then tested again. Making this explicit and intentional would be an effective way of integrating science into the political and policy domain without sacrificing the fidelity of science itself.

16

More than just widgets

Various kinds of fear distort scientific judgements, just as they do other judgements: but the self-deceiving factor seems to be a set of euphorias. The euphoria of gadgets; the euphoria of secrecy. They are usually, but not invariably combined. They are the origin of 90 percent of ill-judged scientific choice.

C.P. Snow[1]

Snow was convinced that politicians do not understand science. A step towards bringing politics closer to science will come with a better description of how science is functionally relevant to society.

Science operates in four main functional modes, each representing a contribution to wider societal needs. The first mode involves the production of what can be called widgets and which C.P. Snow referred to as 'gadgets'. *Widget science* is often motivated by the wish to create new technological solutions. Next is *operational science*, which involves providing direct support for the functions of government itself and many other corporate structures in society. Then there is *foresight science*, which helps us to understand the future. Finally, there is *discovery science*, which is the quest for basic knowledge.

All scientific disciplines have feet in each of these camps, but their presence in each varies greatly. Many scientists like to think they are detached by imagining they are mainly working on discovery science, but when seen through the window of utility for solving problems an immense amount of biology and biotechnology is really widget science, as is much of physics and chemistry. Deep down, the motivation for very little scientific research is likely to be purely disinterested *discovery*.[2] As I suggested in chapter 14, research scientists are really interested in solving problems. A difference may exist, however, in whether research is done mainly for *public good*, where the outputs are connected to non-personal gain, or for *private good*, where the outputs are connected to personal gain.[3] Some science will service both, but being more explicit and realistic about this public/private split might solve some of the dilemmas for scientists and how they are judged. For example, most scientific research has a strong component of private good, not just for those working in the private sector, but also for researchers who are motivated to perform because their salary may be linked to their performance or they may qualify for professional recognition as a result of their performance. I am sure that many scientists would claim that their main motivation is to enhance public good, but most of the systems they work within are geared to providing incentives which are based on private good. I suggest it was lack of clarity about this public/private split which was really at the root of a lot of the negativity about genetically modified foods. Many of the scientists involved failed to explain how the benefits would flow to ordinary people.

Widget science is probably most commonly associated with this drift towards private gain. Widgets are often created by boiling the structure of Nature down to its basic components and building these back up into things which are useful. Statistics, economics and social science, but also a lot of ecology and geology, lie in the domain of operational science, which deals with the realities of Nature in all its huge diversity of form and function. Weather forecasting is an archetypal example of operational science. Operational science is often applied in tactical decision-making and is more firmly grounded in the domain of delivering public good. The third type of science, foresight science is often more strategic and, again, is skewed towards delivering public good. Computational modelling, including climate science, but also an increasingly important field of systems science used to predict or project how critical socially based and Nature-based systems are likely to change in the long-term future, are typical examples of foresight science.

Examples of discovery science would be research done in physics on subatomic particles, the world of the James Webb Telescope plunging back through the depths of time or studies conducted in crystallography which eventually unravelled the structure of the genetic code.

The production of widgets is probably what most people see and understand about the function of science in society as a whole. The term 'widget' covers anything from vaccines to jet engines, smartphones, computers or especially flavoursome tomatoes. Widgets can be very big, like the internet, or very small, like a photonic transistor or a molecular motor engineered as a nanoparticle. They can be formed from multiple small widgets stacked together to make larger, composite widgets like a smartphone. They can be designed to address technical issues like climate change or social issues like equality. Widgets can be wonderful, and when it comes to things like neuromorphic computers, then they might tell us a lot about ourselves too. They could be the solution to many global problems.

Widget science is important politically because it is most often understood as the kind of science which results in innovation which leads to patents and which supports economic growth through market mechanisms – hence its strong connection with the generation of private good.[4] It is what generates tax revenues for governments rather than costs. Most government strategies for science are, therefore, weighted heavily towards this function, focussing in the present on things like advanced materials, informatics, semiconductors, advanced computing, genomics, bioengineering, quantum computing, energy and robotics. These are seen as the lifeblood of progressive economies and an engine house of wealth creation. This is largely the focus of policies specifically targeted at what people want science to do for them, sometimes known as *policy for science.*

But widget science is the *golem* which both befriends and torments the beast in the politics factory. It can be loved most of the time but can come back to bite us. Most widgets have the capacity for *dual use,*[5] for good and for bad. For example, even though widget building does a great deal of good, it tends to feed the supply side of economies by inventing things people want to buy, and in so doing it feeds rampant, unconstrained consumption. As Andy Warhol observed, 'Buying is much more American than thinking,'[6] and widget science is playing to our tendency to want to buy rather than to think.

Increases in production efficiency brought about by innovations in widget science, which is often seen as a superficially positive consequence, is usually matched by increases in consumption. Known as Jevon's paradox,[7] this means widget science does not generally result in us cashing in on improved production efficiency by reducing consumption. For example, science has made light bulbs many times more energy efficient, but instead of this resulting in a reduction in energy consumption, we just use more light bulbs.[8] It is the cost of 'the pursuit of happiness', the words of Thomas Jefferson in the preamble of the US Declaration of Independence. Hedonism means that widgets often designed by scientists for ethically sound reasons, such as conserving natural resources and making people well, end up doing the opposite. Ethical questions buzz around widget science like wasps around a honey pot. Snow thought that it deprived people of a capacity to see the bigger picture. Graeber saw it as the source of spiralling debt as consumers became victims of pressure selling of widgets and thought that it deprived people of their humanity, making them slaves to big business.[9] But widget science can also illustrate an ethical vacuum in politics: the invention of a new drug to treat Alzheimer's disease stimulated one senior political figure to see its potential for reducing the costs of residential care, rather than for alleviating suffering.[10]

Even apparently clever people become intoxicated by the possibilities which arise as a result of widget science, and one result is that the accounting systems used by governments see investment in this specific kind of scientific research in the same way as they see investment in capital infrastructure.[11] It levers science and the rationale for its existence into the strait-jacket of consumptive economics rather than putting it at the service of justice. Knowledge is increasingly viewed as a national (and nationalistic) asset, and large public investment funds are created to ensure that widget science remains a significant part of national economies. In the post-Bretton Woods economic era, kinetic warfare has largely been replaced by economic co-dependency and competition between nations, but this still involves a kind of warfare using weaponry in the form of relative dominance in widget science.

In contrast, operational science occupies a role which has a much lower social value among politicians, the public and the scientific elite. It lacks the patina of greed associated with widget science. It provides technical support for the activities of government and businesses. It provides six-hourly weather forecasts, tells us how much food we produce, where the water we drink comes from and whether it is pure or polluted. It tells us where to find the minerals we need to mine in order to supply the raw materials for almost everything we turn into widgets, and it tells us how much of the reserves of those minerals we have left. It tells us about the rise and fall of the tides, the stocks of fish in the sea, the purity of the air we breathe or the number of viruses there are waiting in animal hosts which could infect people. It also tells us how we use widgets both to make our lives better and to make us mentally and physically ill.

I used a lot of operational science while I was a CSA. This kind of science helps to keep us safe and to stop bad things happening. Much of the science undertaken to help make better decisions during the COVID-19 pandemic was operational. It is used to monitor for disease threats, chemical and radiological hazards and to oil the wheels of the economy by ensuring that the standards we use for weights and measures are maintained and improved. Currency trading would not be possible without the help of operational science which maintains a standard for time so that trades are protected against 'wire fraud'. The list of operational science functions which most of us take for granted is very long indeed.

Operational science can also be described as *science for policy* because it is what makes the policy machinery tick, but it is sometimes demonized by shallow-minded politicians as the science that creates *red tape*, getting in the way of economic objectives because it often underpins regulations designed to create fairness within market systems. The costs of air pollution would be huge if operational science was not there to hold manufacturers to account for the pollution they produced. Government needs this science for policy mainly to stop the burgeoning of wicked problems.

Foresight science has become increasingly important in recent decades. Examples include the projections of the future made by the IPCC and during the COVID-19 pandemic, when it was used to tell politicians what the impact of the disease was likely to be under different policy scenarios. Foresight science also helps us to understand how much planetary boundaries could be stretched or broken by having 10 billion people on the planet by 2050 and what the options are for heading off the worst effects. It often uses knowledge from the past, as well as structural insights about the way systems are formed, to project forward into the future, as is the case with climate projections. These efforts help to provide the context of the future we are stepping into.[12]

The governments of most advanced economies have foresighting research capabilities. In the US, this is known as the Center for Strategic Foresight, and the OECD has a similar function. Many governments use foresighting to help them create a strategic picture of how the world is likely to change in the future. Most finance ministries in countries with advanced economies run financial models to project the response to different budget scenarios, and they increasingly use those models to help design the right interventions to control inflation, unemployment or investment. These models help to take the subjective component out of forecasting by externalizing sociological and political considerations and placing them in their rightful place, at the end of the logic chain, rather than allowing them to interfere to produce the 'right' result. The RAND Corporation, which was created soon after the Second World War, was an early example of a scientific foresighting function and was largely responsible for creating a new form of mathematics known as *game theory*, mainly to help predict how belligerent countries might behave in different circumstances. It is insights from this which have allowed NATO to play a careful game of chicken with Russia over its war with Ukraine.

Foresight science benefits from technical developments in computing and artificial intelligence. It often uses models, sometimes also known as *digital twins*, of the systems being managed. Digital twins are an invention derived from process engineering where the whole engineered system is encoded in a computer, thus allowing an operator to investigate the strengths and weaknesses of a design. The online supermarket Ocado uses a digital twin of its technology which drives down construction and commissioning costs and drives up operational efficiency. Once constructed, the digital twin is updated by real data and 'learns' from the experience, thus gradually becoming better and better as it encounters unpredicted scenarios.

In the context of government, however, the role of foresight science is going to become ever more influential in determining policy. Governments manage systems which have complexities far beyond the capacity of human brains to compute or understand. Foresight science is the field which takes on the challenge of attempting to understand and forecast system behaviour in response to different policies. It therefore has the capacity to downgrade wicked problems to complex problems or even to tame them. It can learn from the experience and so naturally builds in adaptive management of policy implementation.

Sometimes these different kinds of activities in the scientific enterprise work reciprocally. For example, whereas widget science tends to support the supply of private goods, and therefore increase the flow of consumption, foresight science creates the rationale for restraint, for policies which restrict demand, because it highlights the damage done by consumption. Widget science tends to be attractive to politicians because, in very broad terms, it is giving people what they want – it feeds hedonism – whereas foresight science tends to be less popular because it often illustrates forthcoming challenges which are about politicians saying to people they cannot have everything they might want. The ensuing political debate impinges on justice and the rights of people in the present and the future, as illustrated by the foresight science which underpins projections of human-induced climate change. It is impossible to imagine this debate without a significant role for science, at least in setting out the consequences of different pathways into the future, but also for helping people to differentiate between science which mainly supports public rather than private good.

17

Fixing the marriage

At the beginning of this book, I proposed the testing of four ideas: that science is inherently political; that science is currently actively excluded from contemporary political structures; that this exclusion leads to injustice; and that we can do things which will help to fix these problems. In my search for appropriate metaphors, I have compared the solutions with taming a 'beast', because the socio-political axis which benefits from using scientific knowledge has much of the emergent behaviour one could expect from a complex organism, an organism which on the one hand has free will but on the other hand is beset by a bewildering set of constraints it does not understand. Helping the 'beast' leads one into a hopeless cycle of trying to manipulate reality to align with its wishes and expectations rather than trying to convince it to change, to embrace reality rather than the products of its own imagination. The 'beast' builds expectations which are unrealizable, leading many people to feel that they are not getting a fair deal. I have suggested ways of structuring our thinking about how science and politics can be blended to help solve this problem, but because this is not easy, in the pithy words of one focus group participant, 'we always make an arse of it somewhere'. In this final chapter, I want to suggest ways to make less of an 'arse' of it.

I hope by now that the reader will have discarded any idea that science is apolitical and value-neutral or that being political requires scientists to enter the mire of adversarial politics.[1] Nothing could be further from the meaning of this book.

While experience suggests it is generally true that building scientific knowledge into public discourse and policy should, at least in theory, lead to better outcomes, there are important exceptions. These happen especially when scientific uncertainties are not fully revealed, when scientific knowledge is biased or subverted by being pre-shaped by politics, and when the risks from scientific innovation are not built into informed public debate. Nobody should misinterpret this book as a manifesto for technocracy, but justice will be better served if the quality of science is not diluted by its contact with politics.

I have also presented the current relationship between science and politics as being something like a troubled marriage with one dominant partner and inadequate reciprocation. I have explained that there is much that can be done to improve, perhaps even fix, this by understanding how science develops its narratives in ways which can easily be connected into political narratives; by applying discipline when defining problems so that they can be driven towards simplification; by conceptualizing the progression of policy as a learning process homologous with experimental science; and by being clear about the balance between public and private good coming from science because this influences trust. But these kinds of activities need to be hard-wired into the political and bureaucratic process, to be built into the infrastructure of what I have called the politics factory. Experience suggests that a laissez-faire approach to working this out will result in a decline back to the axis of ideology and the corruption of science. Once there, scientists end up being put in metaphorical boxes, politicians play their Machiavellian games and humanity threatens to fall towards a Hobbesian base state. Accepting this as normal, as much of 20th-century political science did, seems to be defeatist. It is not an expression of squishy liberalism to imagine that we could do a lot better. Surely, one of the greatest failures of adversarial politics must be its incapacity to acknowledge the large quantum of shared interests compared with its capacity to amplify and focus on our small quantum of differences.

Anybody who imagines that what I am saying in this book is also a manifesto for the geeks[2] would be mistaken. The interests of science as a profession and endeavour are obviously important and need to be nurtured, but when scientists start to look after their own interests – their private interests rather than focussing on public interests – then science itself becomes subverted. So much that I see of the arguments made in the domain of *policy for science* seems to be in pursuit of science for the sake of scientists rather than science to champion objectivity in public discourse and collective decision-making to create better public policy for the greater public good. When scientists start to look after their own interests for their own sake, then the whole endeavour will have lost its way.

When scientists sound like another bleating faction wanting a greater share of public funding, this is caused partly by the corruption of science as a pathway to justice because of politicization by consequence, coercion or association. The dilemma for scientists is that they are forced to participate within the political ferment when, by participating, they become corrupted, an expression of what I call the *scientifico-political predicament*. Science is made to sing for its supper in ways which undermine its validity and purpose. To pacify critics, scientists then create narratives about how much science is really constructed to help them. That it is a 'powerhouse of economic growth' is a classical pacification tactic playing to those for whom the only way of creating public good is through private wealth creation.

How, therefore, in a structural sense, can science be appropriately integrated into democracies to deal with this dilemma while avoiding the subjective by-pass and gravitation towards technocracy or plutocracy?

Journalistic freedoms are one of the foundations of democracy, and I suggest it is reasonable to think of scientific freedoms, like journalistic freedoms, as an off-shoot of the freedom of speech. I suspect there is a broad correlation among societies between support for scientific and journalistic freedoms. But the institutional guardians of freedom of the press and freedom of science differ greatly in their capacity to rebuff political meddling. I have seen through my various roles in the governance of science just how much politicians meddle in scientific institutions, and just how weak scientific institutions are at resisting.[3] There is nothing of the robustness with which journalists guard their independence. When the UK left the European Union, the scientific enterprise became a casualty of the ensuing political mess, but it was an accepted norm that press freedoms would never be affected. Politicians know there is a red line concerning journalistic freedoms, but there are no red lines for science.

The rise of scientific power in the second half of the 20th century created competition between nations for scientific prowess. The idea of 'open', freely shared scientific knowledge therefore has become anathema to many nationalistic politicians. The US has been preoccupied by China's accrual of benefits from research funded by the US taxpayer. The complaint from the US is that there is insufficient reciprocation on the part of China to justify US-funded research benefits leaking to China. It is seen as a silent leakage of the capital value of research from one owner to another, a very capitalist and utilitarian view of the purpose of science, and the old, reliable chestnut of national security is used as a pretext for undermining the openness of science. For the egalitarian ideals of science to be fully realized, trust is essential. Trust flows from reciprocation and trust has been lost in this case.

Some of the existing bureaucratic hardware in the politics factory includes the *Madisonian* model of democracy involving three strands of government – the legislature, the judiciary and the executive – operating supposedly independently and in balance. How should science play its role within this framework?

Currently, *policy for science* and *science for policy* are largely under the control of the executive branch of government, but this is what mostly results in corruption and the dilution of scientific freedoms. Nobody in functional democracies would ever suggest that control of the press would sit entirely within the executive function of government – that is the preserve of totalitarianism. As we have seen with the example of how air quality has been dealt with, the judiciary sometimes intervenes to correct the executive branch to take a more science-based way of thinking, but, depending on the judicial system and the subject, judges can be just as quixotic as politicians. The legislature can be more proactive, but is populated mostly by the same kind of people who sit within the executive and is quite prone to ignoring science itself.

There is something, therefore, not quite right about these governance structures, which seem to come straight out of the playbook of pre-Enlightenment government, when there was little appreciation for the Laws of Nature and when civil or common law dominated. I have previously explained the existence of differences between legal and scientific definitions of 'evidence' and also how the 'troubled marriage' of science and politics has been attributed to a consequence of two very different understandings of truth. If we persist with the false notion that legalistic truth trumps scientific truth, as is the case within our common and civil law systems of governance currently in place, then we will continually struggle to match aspiration to reality, to align our ways of living so that they respect the constraints of Nature. We need a much better way to integrate and align the tellurian laws holding together the fabric of human society with the Laws of Nature.

Some parts of the infrastructure for achieving this already exist and provide an example of how we could proceed in future. *Official statistics* are a genuine attempt to ground the self-interested decision-making in the politics factory in real rather than bogus information. Strong safeguards exist to stop political interference in these statistics from any branch of the democratic structure, and often the 'statistical offices' within nations can hold to account those who misuse statistics, including politicians. These statistics provide information about everything from crop production to hospital performance and immigration, from the quality of education to economic production. They are recent innovations, but are now so important to the smooth running of society that the United Nations has mandated how nations should govern and generate them.

In the UK, the authority which oversees this 'has the statutory objective of promoting and safeguarding the production and publication of official statistics that "serve the public good"'[4] and its leadership is appointed by the head of state. A similar process has been developed for oversight of analysis of national budgets: in the UK this is known as the Office for Budget Responsibility, which exists under a charter established in law. It exists 'to examine and report on the sustainability of the public finances'.[5] Politicians who have in the past tried to ignore it have been made to regret doing so by the reactions of the financial markets. This is not quite as secure from political interference as the statistics authority, but has many more protections than the current system through which science is delivered and governed.

These are cases showing how the overall governance of the quest for rational, objective analysis – statements which reflect our best understanding of reality – can be exterior to the domain of rhetorical politics. They sustain accountability within all three branches of government and, if implemented for science, they would oblige them to listen to what science says, whether this is about the flow of the solar wind, the structure and function of the atmosphere and ocean, the emergence and prevalence of disease, the safety and supply of food and water, the evolution of national risks or any other aspect of our technologically dominated lives. Any deviance from the emerging technocratic advice would need to be justified and debated, a process for which both a free press and free science are vital components. The free flow of scientific information from a politically empowered but ideologically neutral scientific community, with appropriate signalling as to the quality of that information, would help to create a foundation for useful value-based arguments which rightly have a place within adversarial debate.

However, much of the infrastructure needed to achieve this is absent. Instead, as things stand, much of *science for policy* is serviced by an *ad hoc* network of science advisers and scientists, including economists and social researchers, who are embedded within government departments and are working only for the executive function of government, something which is clearly a corrupt model because of all the conflicts of interest which exist. These can be represented by a few individuals working at senior level as well as large institutional structures of various kinds but which largely exist to service the scientific needs of executive government. All of these have almost no effective protection from political interference.

Examples of how poorly these perform can be illustrated by the ways in which economics plays into policy, and I have already provided examples of this earlier in the book. Governments normally work to a set of guidelines for assessing the viability of new policies overseen by economists. These produce *impact assessments* which consider the balance of costs and the potential benefits of using scientific knowledge, but they are rarely honest about the uncertainties which underlie their conclusions. They are an example of confusion between legalistic and scientific doctrines of evidence where both are mixed up and used selectively in ways which predetermine the outcomes of assessments based on the preferences of those involved in doing the assessments. They are subjectivity deceptively dressed up as objectivity. They can be made to swing towards support for the political preference of the political masters or towards the wielding of technocratic power, that dark place where economists themselves are running the politics factory by stealth.

Certainly, when I was a scientific adviser working within the executive structures of government, I felt coercion to conform to the current political will, but I also saw into that dark space where I could control policies by tilting the 'evidence' (and therefore the policies) to match my own preferences. Scientific advisers like me are threaded through the executive side of governments, but we walk a tightrope which is loaded with jeopardy.[6] This makes the system of Chief Scientific Advisers an inadequate system for ensuring that the systems of natural law, including the Laws of Nature, are appropriately represented within the governance of nations.

A much more satisfactory system would be, therefore, for scientists who find themselves working within the politics factory to be responsible to an authority (or authorities) which is external to the executive, legislature and judiciary, possibly appointed by the head of state, in the style of current statistical authorities. In nations where that head of state is the *de facto* head of the executive branch, there would also need to be assurance that there is endorsement from all three branches of government. Scientists operating in any sense within the politics factory would need to be formally registered as professionals whose qualifications and conduct were subject to quality control assessment and accountability. This would also make 'experts' recognizable, in the same kind of way as for practising doctors and lawyers, leading to greater confidence in their authenticity. This 'authority' would also be required to provide oversight of the institutions which conduct scientific research in the public interest, including places like universities and those which currently sit completely within the executive branch of government, guaranteeing and guarding their quality and independence from external influences, and especially from politics.

A statutory 'authority' of this type would be required to operate and enforce a system of internationally accepted standards. It would do what Bob May tried and failed to achieve: namely to set out and ensure compliance with a set of rules about how politicians and the wider players in the politics factory need to use science, and to call them to account when they depart from those rules. In broad terms, those rules would say that you follow the guidance of science but it is for you to formally state your reasoning if you choose to depart from that guidance. This opens those within the public sphere to scrutiny about whether they have used science appropriately, and if not, why not?

To counter any drift towards technocracy, such a scientific authority would need to champion 'open' science and scientific freedoms, in the same way as journalistic freedoms are championed. Eventually, having a scientist in the room – whether a court, ministerial office or legislative chamber – when decisions are being made should be no different to having a lawyer present when the situation dictates it.

An alternative system for the structural inclusion of science in democratic processes has been suggested.[7] This involves adding to the current capabilities of the legislative arm of the Madisonian model by replacing the bicameral system of legislatures (involving having lower and upper legislative chambers) with a tricameral system involving the addition of a third chamber containing scientific and technical experts. The upper house in the UK, the House of Lords, has some elements of such a system because it has members known as 'cross-benchers' who are appointed because of their technical knowledge. An enhanced version of such a system might work, but it would certainly add to the already substantial inertia within legislative processes and would most probably end up with its members aligning along party-political lines, leading to yet more ideological rancour. It would also do nothing to increase the presence of scientific realism within the other two arms of the governance of nations.

In my final words on this subject, I return to philosophy. The political philosopher John Rawls saw the development of reciprocal, cooperative relationships as a generally beneficial process leading to fairness forming a cornerstone of justice.[8] He also developed an *idea of public reason* which concerns, among other things, how scientific claims are incorporated into political argument.[9] What I have described in this book is how the current *ad hoc* process of incorporation of science into public reason is insufficient for the modern age, and it may even be pernicious. It often leads to perversities and sometimes even to precisely the opposite of the enhancement of justice. When we get the balance wrong between the purpose of science for generating public versus private benefits, then injustice flows.

Rawls and others (like Isaiah Berlin) argued that there is a conflict between equality and liberty. For equality to apply, we need to abide by rules to stop individuals accruing private goods at the expense of others and this constrains liberty. The public or private good generated by scientific research is closely aligned respectively with the axes of equality and liberty: public good has a lot to do with sharing and equality, whereas private good has a lot to do with liberty to use, exploit and own. How we, therefore, deploy science across society to achieve an appropriate balance between equality and liberty says something meaningful about the issue of how we perceive the delivery of justice. The pacification tactic of seeing science simply as a 'powerhouse of economic growth' rather than as tool for supporting the health and welfare of people and the planet could, arguably, be deploying science to undermine justice.

But perhaps a more powerful way in which science acts as a pathway to justice is its potential to neutralize the trade-off between equality and liberty. It has the capacity to rotate the axis of orientation in how values flow through society. Although we tend to distil value down to exchangeable currencies like money, debt and credit, we increasingly understand that knowledge itself has value. The free exchange of scientific knowledge has got much to do with the creation of equality because value from knowledge flows in ways which money cannot. For example, open and shared scientific knowledge allowing the world's poorest people to pull themselves up by their own bootstraps is much more likely to work than debt peonage.[10] Therefore, liberty in the form of knowledge-sharing likely leads to equality.

Perhaps, then, science has the key to solving the equality–liberty trade-off. This is a highly contested space and it
is why science is such a politically significant subject. One
can fume, as C.P. Snow did, about the exclusion of science
from the domain of public policy. One can also explain, as
I hope I have done in this book, the mechanics and sociology of why this happens and offer up ways of fixing the
problem. But the fundamental reason for exclusion, subversion, perversion and corruption is that science is a threat
to old ideas and old hegemonies promoted even today by
the priests, shamans and princes – the charlatans – who
still have dominant voices within the politics factory. The
easy route would be to leave 'science to the scientists and
politics to the politicians', but this will never successfully
address the *scientifico-political predicament* and the world
will be a much worse place as a result. Scientists hold the
key to ridding the world of hocus-pocus, but if they demur,
then there is nothing to stop its spread within politics and,
by default, scientists will be guilty of promoting injustice.

Notes

Preface

1 See S.J. Gould, 'The exaptive excellence of spandrels as a term and prototype', *Philosophical Transactions of the American Academy of Science* 94 (1997): 10750–10755, doi: 10.1073/pnas.94.20.107.

2 The refences to 'princes' is in the sense referred to by Niccolò Machiavelli in his book *The Prince* (1532). It includes dictators and absolute monarchs.

3 Of the great schools of Western philosophy, stoicism is most often associated with the virtues of alignment with Nature, and many eastern philosophies are also rooted in a similar conceptualization. However, in the context of this book I do not suggest that science has a particular affiliation with any one of these schools. The Peripatetic school of Aristotle has much to say about deductive logic, Platonism about inductive logic and Epicureanism about the importance of the fundamental, atomized or quantum structure of reality. All have their place and all are valid ethical and moral perspectives used in modern science, as different ways to grip the twisted topology of reality.

4 Social contract theory, broadly stated, suggests that while people are free to live as they see fit, they also have obligations

to wider society represented by authorities like governments which act to represent the collective good.

5 There is, of course, a rich philosophical literature involving the likes of Hobbes, Locke, Bentham, Hume, Rousseau and Mill, all of whom had much to say on the subject of the balancing of individual interests against the common interest. Much of this philosophy is relevant to this book, but it is beyond its scope and so must remain implicit.

6 R. Carson, *Silent Spring*, Boston: Houghton Miffin, 1962, p. 297.

7 See M. Henderson, *The Geek Manifesto: Why Science Matters*, London: Bantam Press, 2012.

8 Richard Lewontin described 'biology as ideology' in his book of that name (R.C. Lewontin, *Biology as Ideology: The Doctrine of DNA*, Toronto: House of Anansi Press, 1991).

9 J.M. Ziman, *Public Knowledge: An Essay Concerning the Social Dimensions of Science*, Cambridge: Cambridge University Press, 1974, p. 1.

Introduction: The scientific predicament

1 K. Sanderson, 'How to convince a politician: a science adviser's lessons from the pandemic', *Nature* 616 (2023): 20, doi: 10.1038/d41586-023-00938-w.

2 There are numerous examples of people making these statements, but one senior scientist thought that 'creating a firewall that enables science to be science and politics to be politics' would be desirable (N. Subbaraman, 'After Trump, US researchers urge Biden to block political meddling in science', *Nature* 596 (2021): 174–175, doi: 10.1038/d41586-021-02132-2). Another typical example is from the science writer and psychologist Stuart Ritchie ('Science and politics shouldn't mix – as scientific journal Nature has discovered', *iNews*, 24 March 2023, https://inews.co.uk/opinion/science-and-politics-shouldnt-mix-as-scientific-journal-nature-has-discovered-2229379.

3 It is for this reason that I suggest the frequent reference to 'decision-makers' within the science policy literature is wrong. Within democratic systems, we are all decision-makers.

4 Solly Zuckerman's view was that 'there is nothing to stop chief scientific advisers from changing roles and from entering the race which ends in leadership and power' (S. Zuckerman, 'Science advisers and scientific advisers', *Proceedings of the American Philosophical Society* 124(4) (1980): 253).

5 As a member of various governing boards across government, charity and commerce, I see the democratic process built into the fabric of society.

6 L. Strauss, 'Strauss' responses to contemporary defenses of the fact–value distinction', in C.H. Zuckert (ed.), *Leo Strauss on Political Philosophy: Responding to the Challenge of Positivism and Historicism*, Chicago: University of Chicago Press, 2018, pp. 96–120.

7 A system championed by theorists like Charles Merriam and Quincy Wright at the Chicago School.

8 M. Weber, 'Politics as a vocation', in H.H. Gerth and C.W. Mills (eds), *From Max Weber: Essays in Sociology*, Abingdon, Oxon: Routledge, 1991, p. 128.

9 See B.B. Palmer, '4 science-based strategies to tame angry political debate and encourage tolerance', *The Conversation*, 27 February 2020, https://theconversation.com/4-science-based-strategies-to-tame-angry-political-debate-and-encourage-tolerance-130984.

10 H.J. Morgenthau, 'The evil of politics and the politics of evil', *Ethics* 56(1) (1945): 1, doi: 10.1086/290471.

11 Morgenthau thought that ethics was 'indistinguishable from science' (H.J. Morgenthau, *Scientific Man vs. Power Politics*, London: Latimer House Ltd, 1947, p. 146). For more detailed discussion, see B. Gewen, *The Inevitability of Tragedy: Henry Kissinger and His World*, New York: W.W. Norton, 2020, chapter 4.

12   L. Strauss, *The Political Philosophy of Hobbes: Its Basis and Its Genesis*, trans. E.M. Sinclair, Oxford: Clarendon Press, 1936.

13   H. Boytchev, 'Keep science and politics separate, say German professors', *Research Professional News*, 18 April 2023, https://www.researchprofessionalnews.com/rr-news-euro pe-germany-2023-4-keep-science-and-politics-separate-say -german-professors/.

14   R.A. Pielke, *The Honest Broker: Making Sense of Science in Policy and Politics*, Cambridge: Cambridge University Press, 2014.

15   *The Republic* Book 1, 347c.

16   J. Sinclair, 'Free up science to combat misinformation, top adviser says', *Research Professional News*, 20 March 2024, https://www.researchprofessionalnews.com/rr-news-new-ze aland-2024-3-free-up-science-to-combat-misinformation -top-adviser-says/.

17   In 2023, the UK's National Risk Register considered that the probability of another pandemic over the following five years was between 1:20 and 1:4.

18   N. Petrosillo et al., 'COVID-19, SARS and MERS: are they closely related?', *Clinical Microbiology and Infection* 26 (2020): 729–734, doi: 10.1016/j.cmi.2020.03.026.

19   Albert Einstein said this when he was confronted by the apparent randomness in Nature brought about by quantum processes.

20   They were also evident in the UK after its decision to leave the European Union.

21   Most current agricultural research is focussed on increasing the productivity and sustainability of traditional agricultural systems. This has led to small, incremental improvements over the past thirty years but the rate of innovation is slowing. It is likely that the only viable solution to supplying sufficient food to people without also ruining the biodiversity of the planet and exhausting soils is to build very large factories

linked to a significant sustainable energy source. This would take in carbon dioxide and a few other basic elements and turn them into basic food commodities (carbohydrates, proteins and fats).

22 These toxic chemicals, many of which accumulate within the environment, are highly resistant to natural breakdown. They are almost universal components of various common materials.

23 Since I live in Scotland, these numbers are for the average person living there. See *Delivering Scotland's Circular Economy*, https://www.gov.scot/binaries/content/documents/govscot/publications/consultation-paper/2022/05/delivering-scotlands-circular-economy-consultation-proposals-circular-economy-bill/documents/delivering-scotlands-circular-econ omy-consultation-proposals-circular-economy-bill/deliver ing-scotlands-circular-economy-consultation-proposals-circu lar-economy-bill/govscot%3Adocument/delivering-scotlands-circular-economy-consultation-proposals-circular-economy-bill.pdf.

24 In 2022, the US processed more than 10 billion livestock animals and, like all other countries, sustained rampantly poor biodefence (A. Linder and D. Jamieson, 'Blind spots in biodefense', *Science* 379 (2023): 621, doi:10.1126/Science.adg 9237).

25 Government risk assessments often find that the risk of a particular disaster is relatively low, but they also often fail to recognize that the more risks there are, then the aggregate probability of something bad happening increases in proportion to the number of those risks.

26 Boiling down what science does for society to metrics that bean-counters recognize is popular, even though what we value is so much more than money. It seems that every £1 invested in medical research delivers a return equivalent to around 25p every year, for ever! (*Investing in UK R&D 2019*, The Royal Society, London, https://royalsociety.org/-/media

/policy/projects/investing-in-uk-r-and-d/2019/investing-in
-UK-r-and-d-may-2019.pdf).

27 Natural realism is also often associated with the 'common
    sense' philosophy of the 18th-century Scottish Enlightenment
    and is often attributed to Thomas Reid (1710–1796). In the
    sense used in this book, it builds on the Bayesian inferential
    method used to understand the relationship between cause
    and effect. This is a specific acknowledgement that the future
    can only be inferred approximately from knowledge of the
    present and the past, and that this reflects how most organ-
    isms, including humans, deal with the complexities and
    uncertainties of the world around them.

Chapter 1 Beyond two cultures

1 S. Zuckerman, *Beyond the Ivory Tower: The Frontiers of Public
   and Private Science*, London: Weidenfeld & Nicolson, 1970,
   p. 100.

2 Report of a speech made by Sir Patrick Vallance in July 2022
   at the launch of the University of Oxford's Pandemic Sciences
   Institute: S. Inge, 'Vallance: "stay in your tramlines" for science advice and policy', *Research Professional News*, 8 July
   2022, https://www.researchprofessional.com/0/rr/news/uk
   /politics/2022/7/Vallance---stay-in-your-tramlines--for-sci
   ence-advice-and-policy.html#sthash.Riic7fpf.dpuf).

3 IPSOS Mori Veracity Index 2022, https://www.ipsos.com/en
   -uk/ipsos-veracity-index-2022.

4 C.P. Snow, *The Two Cultures* (with an Introduction by
   S. Collini), Cambridge: Cambridge University Press, 2012
   [1959].

5 There are many examples of the apparent exclusion of science:
   (a) The British politician Jacob Rees-Mogg published a book
        in 2019 describing twelve great Victorians who shaped
        the world around them (J. Rees-Mogg, *The Victorians*,
        London: Penguin Books, 2019). None that he chose was
        a scientist or engineer.

(b) In a book titled *Imagine a Country*, ninety-eight Scottish 'thinkers' were asked to describe their ideal future country (V. McDermid and J. Sharp (eds), *Imagine a Country: Ideas for a Better Future*, Edinburgh: Canongate, 2020). The editors, whose aim was to include 'a cross-section of Scottish society' (p. 1), could find space for only one scientist and one economist to express their views among the ninety-six others, who were mostly from across the arts and humanities.

(c) Dorothy Byrne, former Head of News and Current Affairs at the UK television station Channel 4, noted at a seminar about journalism during the COVID-19 pandemic in June 2020 that most of those reporting on, or deciding about, the conduct of the management of the pandemic came from a very narrow intellectual stable of people who had graduated in fields like history, philosophy, law and politics (D. Byrne, 'What journalists can learn from their mistakes during the pandemic', Reuters Institute/University of Oxford, 17 June 2020, https://reutersinstitute.politics.ox.ac.uk/risj-review/what-journalists-can-learn-their-mistakes-during-pandemic).

(d) This was a view backed up by Kate Bingham, a venture capitalist, who led the scaling-up process of delivering one of the first COVID-19 vaccines to be used (K. Bingham, 'Britain is driving away innovators in life sciences', *The Times*, 22 November 2021, https://www.thetimes.co.uk/article/britain-is-driving-away-innovators-in-life-sciences-c6b0xrpxm).

6 As of April 2024, Britain has had seventeen post-war Prime Ministers. Three never went to university, one went to Edinburgh University and the remaining thirteen all went to Oxford University. Only one, Margaret Thatcher, had a science degree..

7 In 2019, the UK Parliament's lower legislative chamber had 650 members but only one (0.15 per cent) had a science Ph.D.

and only about half of them could answer a basic question in statistics ('New RSS survey tests statistical skill of MPS', Royal Statistical Society, 11 February 2022, https://rss.org.uk/news -publication/news-publications/2022/general-news/new-rss -survey-tests-statistical-skills-of-mps/). Perhaps a little over 0.5 per cent of the adult population in the UK have a science Ph.D., so it appears science is seriously under-represented in the democratic system.

At the level of the United Nations, science representation was completely forgotten until it was embedded as the 'S' within UNESCO, but only as an afterthought ('UNESCO must reform to stay relevant – and reconnect people through science', *Nature* 587 (2020): 521–522, doi: 10.1038/ d41586-020-03311-3).

In 2019, 27 per cent of all higher education degrees in the US were in science and engineering ('The state of US science and engineering 2022', National Science Board Science & Engineering Indicators, https://ncses.nsf.gov/pubs/nsb2 0221/u-s-and-global-stem-education-and-labor-force). This intellectual diversity was not repeated throughout the executive and the legislature. Almost all of these legislators had degrees but, in the 116th Congress, which set a record high for inclusion of scientists and engineers, only 5 per cent had degrees in science or engineering and most of those who did were engineers. In contrast, 27 per cent had law degrees and yet the number of law graduates in the US is about one-third of those from science and engineering (J.A. Nekuda Malik, 'US 116th Congress sets new record for members with STEM backgrounds', *MRS Bulletin* 44 (2019): 88–90, doi: 10.1557/ mrs.2019.28; Membership of the 116th Congress: A Profile. Congressional Research Services, updated 17 December 2020).

8 R. Lomas, *Freemasonry and the Birth of Modern Science*, New York: Barnes & Noble, 2004.

9 For example, whereas the measurement of time was once

an internalized and aesthetically sensed part of our lives, the invention and refinement of the clock using an oscillator of some sort has gradually allowed time to become progressively more available and refined, enabling time to be commoditized. It now enables present-day financial markets and the Global Positioning System to function. We have become obsessed by time and we talk about losing time, spending time, saving time or even killing time, but this has a sinister side involving the systematic use of labour. We 'clock on' and 'clock off' and our time is bought and sold for money, something which the anthropologist David Graeber called 'spiritual violence' (D. Graeber, *Bullshit Jobs: A Theory*, London: Allen Lane, 2018). Time has had aesthetically defined meanings such as those proposed by the French philosopher Paul Ricoeur, who saw human time as the narrative we create about ourselves and which defines who we are (P. Ricoeur, *Time and Narrative*, 3 vols, trans. K. McLaughlin and David Pellauer, Chicago: University of Chicago Press, 1984–1988). The question of who owns personalized time and how it is deployed is at the root of much political aggravation and philosophy. This experience of how science changes our lives possibly points to how the new revolution in artificial intelligence is going to affect us too. For a fuller description of the sociological changes associated with the measurement of time, see E.P. Thompson, 'Time, work-discipline, and industrial capitalism', *Past & Present* 38 (1967): 56–97, http://www.jstor.org /stable/649749.

10 See D.J. Chalmers, *Reality +: Virtual Worlds and the Problems of Philosophy*, New York: Norton, 2022.

11 See, for example, the case of the novelist who 'faked' the death of her online persona (C. Bailey, 'Fan outrage at Susan Meachen, the romance novelist accused of faking her death', BBC News, 8 January 2023, https://www.bbc.co.uk/news/ world-us-canada-64179548).

12 R. Lewontin, *The Triple Helix: Gene, Organism and*

Environment, Cambridge, MA: Harvard University Press, 2000.

13 M. Warman, 'Stephen Hawking tells Google "philosophy is dead"', *The Telegraph*, 17 May 2011, https://www.telegraph .co.uk/technology/google/8520033/Stephen-Hawking-tells -Google-philosophy-is-dead.html.

14 https://www.quora.com/What-does-Peter-Singer-think-of -Stephen-Hawkings-2011-proclamation-that-philosophy-is -dead.

15 C.P. Snow, *Science and Government*, Cambridge, MA: Harvard University Press, 2013 [1960].

16 In a cutting analysis of the 'British system' published in 1962, Don K. Price, who was the Dean of Public Administration at Harvard University, said that 'the Administrative Class . . . is still dominated by men trained in the classics and historical studies; not one in twenty among these guardians of public policy has a scientific or technical education' (D.K. Price, 'The scientific establishment', *Science* 136 (1962): 1100).

France has had its prestigious École Nationale d'Adminstration, or ENA, which trains very select groups of individuals to be senior government administrators. Some become senior politicians too. The great majority of the annual intake of eighty to ninety students have parents who occupy senior government or business positions. This system was also rolled out over the old French empire. 'Énarques', as the graduates are known, receive no scientific education; knowledge of medicine, engineering and science and technology appears to be extra-curricular.

In the US, a similar elite exists, where the Ivy League universities are predominant within politics, business and finance (C. Burgat and T. Billing, 'Everything you've ever wanted to know about educational pipelines to Capitol Hill', LegBranch.ORG, 4 November 2019, https://www.legbran ch.org/everything-youve-ever-wanted-to-know-about-edu cational-pipelines-to-capitol-hill/). Price also notes in 'The scientific establishment' that the US has a much more fluid system of skills and knowledge sharing between its public and private institutions than elsewhere.

Among the very few national leaders who have been scientists, Jimmy Carter (US President, 1977–1981) probably qualifies because of his experience as a nuclear engineer. Otherwise there is Margaret Thatcher (UK Prime Minister, 1979–1990, chemistry), Angela Merkel (German Chancellor, 2005–2021, quantum chemistry) and Nikolai Denkov (Bulgarian Prime Minister, 2023–2024, chemistry).

17 Medawar's view is that the person of quality in science had to come from the right scientific stable, which was rich in Nobel Prizes, almost entirely racially white, European or North American and male (P. Medawar, *The Threat and the Glory: Reflections on Science and Scientists*, Oxford: Oxford University Press, 1991).

18 See note 1, this chapter, above.

19 Vannevar Bush perhaps can be credited with inventing the role of science adviser to the top political brass (G.P. Zachary, *Endless Frontier: Vannevar Bush, Engineer of the American Century*, New York: Free Press, 1997).

20 Roger Pielke examined how science works as a social construct when faced with different kinds of politically or policy-related questions (R.A. Pielke, *The Honest Broker: Making Sense of Science in Policy and Politics*, Cambridge: Cambridge University Press, 2014). Susan Owens took a similar approach when describing the rise and fall of the UK's Royal Commission on Environmental Pollution (S. Owens, *Knowledge, Policy, and Expertise: The UK Royal Commission on Environmental Pollution 1970–2011*, Oxford: Oxford University Press, 2015). Both suggested different models, or modes, of behaviour that scientists could adopt in specific circumstances.

21 Peter Gluckman, former Chief Scientific Adviser in New Zealand, illustrated this as a general principle (P. Gluckman,

'Policy: the art of science advice to government', *Nature* 507 (2014): 163–165, doi:10.1038/507163a).

22 See H. Nowotny et al., *Re-thinking Science: Knowledge and the Public in an Age of Uncertainty*, Cambridge: Polity, 2010.

23 In 2014, the United Nations programme set up to consider biodiversity loss (IPBES) accepted the inclusion of 'indigenous and local knowledge systems' as a valid source of knowledge.

24 See UNESCO's report on 'open science' and its benefits (*Open Science Outlook 1: Status and Trends Around the World*, https://unesdoc.unesco.org/ark:/48223/pf0000387324).

25 See his podcast on the BBC, *The Long History of Argument*, https://www.bbc.co.uk/programmes/m0019jxn.

26 In general terms, the skills needed to run large complex organizations might include: a sharp mind; an appreciation of tactical versus strategic interventions; a sense of fairness; perception about the level of honesty in other people and of their abilities; an ability to engage with technical detail; a capacity to recognize, deal with and expose risk, uncertainty and ambiguity; a capacity to simplify rather than complicate; and a quick sense of what motivates people.

27 I served five Secretaries of State, and if my sample is representative, then barely one in five of those who sit in the UK Cabinet are competent to do the jobs they are given.

Chapter 2 The anatomy of a troubled marriage

1 Editorial, *British Medical Journal*, 13 November 2020, doi:10.1136/bmj.m4425.

2 'Kazuo Ishiguro and Venki Ramakrishnan: imagining a new humanity', *Financial Times*, 26 March 2021, https://www.ft.com/content/eca7988d-2961-4b27-9368-ff58c966e969.

3 Quote from Kenneth Bernard, former Assistant Surgeon General to Presidents Bill Clinton and George W. Bush (K. Bernard, 'Biden needs scientists with policy chops', *Nature* 602 (2022): 363, doi: 10.1038/d41586-022-00411-0).

4 O. Houck, 'Tales from a troubled marriage: science and law

in environmental policy', *Science* 302 (2023): 1926–1929, doi: 10.1126/science.1093758.

5 I. Mitchell, *The Justice Factory: Can the Rule of Law Survive in 21st Century Scotland*, 2nd edn, Campbeltown: Burnside Studios, 2020.

6 In 2017, California became the first state in the US to list glyphosate, a common weedkiller, as a potential cause of cancer. Scientific support for this decision was very weak yet the law had to come down on one side or the other. If enough cases are brought to court, then multiple jeopardy means that somewhere and at some time the judgment will fall in favour of the plaintiffs.

7 R. Carson, *Silent Spring*, Boston: Houghton Mifflin, 1962.

8 J. Vidal, 'Norman Borlaug: humanitarian hero or menace to society?', *The Guardian*, 1 April 2014, https://www.theguar dian.com/global-development/poverty-matters/2014/apr/01 /norman-borlaug-humanitarian-hero-menace-society.

9 One caveat to this role of genetics is the emergence more recently of a better understanding of epigenetics, where characteristics can be inherited in ways which are not strictly encoded in nuclear DNA and which can be affected by the environment.

10 S. Keen, 'The Soviet era's deadliest scientist is regaining popu- larity in Russia', *The Atlantic*, 19 December 2017, https:// www.theatlantic.com/science/archive/2017/12/trofim-lysen ko-soviet-union-russia/548786/.

11 Q. Schiermeier, 'Russian science: the battle for Russia's brains', *Nature* 449 (2007): 524–527, doi: 10.1038/449524a.

12 W.I. Kolchinsky et al. 'Russia's new Lysenkoism', *Current Biology* 27 (2017): PR1042-R1047, doi: 10.1016/j.cub.2017. 07.045.

13 The quotations are from L.J. Lear, 'Rachel Carson's "Silent Spring"', *Environmental History Review* 17 (1993): 23–48, doi: 10.2307/3984849.

14 N.E. Borlaug, 'Mankind and civilization at another crossroads',

McDougall Memorial Lecture, Sixteenth Session of the Food and Agriculture Organization of the United Nations, 1971, p. 45.

15 D. Meadows et al., *Limits to Growth*, New York: Universe Books, 1972.

16 See N. Rich, *Losing Earth: A Recent History*, New York: Farrar, Straus and Giroux, 2019.

17 I. Sample, 'Scientists prepare for protest: "the march should be a starting point"', *The Guardian*, 21 April 2017, https://www.theguardian.com/science/2017/apr/21/scientists-prepa re-for-protest-the-march-should-be-a-starting-point-march -for-science.

18 P. Weston, 'The rise of eco-anxiety and how to come to terms with climate change', *The Independent*, 22 June 2019, https://www.independent.co.uk/environment/climate-change-exti nction-eco-anxiety-ice-melting-sea-level-wildfire-a8968011 .html.

19 From G.P. Zachary, *Endless Frontier: Vannevar Bush, Engineer of the American Century*, New York: Free Press, 1997, p. 142.

20 D.K. Price, 'The scientific establishment', *Science* 136 (1962): 1099–1106.

21 S. Jasanoff, *The Ethics of Invention: Technology and the Human Future*, New York: W.W. Norton, 2016.

22 Annual Report of the Government Chief Scientific Adviser, *Innovation: Managing Risk, Not Avoiding It*, London: The Government Office for Science, 2014, https://assets.publish ing.service.gov.uk/media/5a7dccf2ed915d2ac884da1b/14-11 90a-innovation-managing-risk-report.pdf.

23 H.M. Collins and T. Pinch, *The Golem: What You Should Know About Science*, Cambridge: Cambridge University Press, 2012.

24 European Environment Agency, *Late Lessons from Early Warnings: Science, Precaution, Innovation*, EEA Report No. 1/2013, https://www.eea.europa.eu/publications/late-lessons -2.

25   H.H. Thorpe, 'Persuasive words are not enough', *Science* 368 (2020): 1405, doi: 10.1126/science.abd4085.

26   Political science sees conspiracy theories as having three anti-science features: (1) nothing happens by accident; (2) everything is connected; and (3) nothing is as it seems (M. Butter and P. Knight, 'The history of conspiracy theory research: a review and commentary', in J.E. Uschinski (ed.), *Conspiracy Theories and the People Who Believe Them*, Oxford: Oxford University Press, 2018, pp. 33–46).

27   See the descriptions of the discoveries of gene drives in M. Cobb, *The Genetic Age: Our Perilous Quest to Edit Life*, London: Profile Press, 2022.

28   M.N. Meyer et al., 'Wrestling with social and behavioral genomics: risks, potential benefits, and ethical responsibility', in E. Parens and M.N. Meyer (eds), *The Ethical Implications for Social and Behavioral Genomics, Hastings Center Report* 53(2) (2023): S2–S49, doi: 10.1002/hast.1477.

29   Generally attributed to the physicist David Mermin when referring to the strange metaphysics of quantum mechanics. It has been expanded to refer to the muzzling of scientists who think too deeply about the meaning and implications of what they are doing (D. Kaiser, 'History: shut up and calculate!', *Nature* 505 (2014): 153–155, doi: 10.1038/505153a).

30   H.H. Thorpe, 'Time to unfriend Facebook?', *Science* 374 (2021): 9, doi: 10.1126/science.abm5515.

Chapter 3 Inside the politics factory

1   P.B. Medawar, *The Art of the Soluble*, London: Methuen & Co. Ltd, 1968, p. 35.

2   B. Gewen, *The Inevitability of Tragedy: Henry Kissinger and His World*, New York: W.W. Norton, 2020.

3   C. Craig, *How Does Government Listen to Scientists?*, Cham: Palgrave Macmillan, 2019, p. 7.

4   There have been considerable efforts to find a vaccine, but this has largely failed so far for both *Mycobactrium bovis*

and *M. tuberculosis*, the form of TB which mainly infects people.

5 A study suggested that most of the cause of transmission of bovine TB was the result of the movement of cattle between farms (E. Brooks-Pollock et al., 'A dynamic model of bovine tuberculosis spread and control in Great Britain', *Nature* 511 (2014): 228–231, doi: 10.1038/nature13529).

6 The Randomized Badger Culling Trial was carried out between 1999 and 2006 and it set the tone for the debate about the management of bovine TB until the present day (C.A. Donnelly et al., 'Impacts of widespread badger culling on cattle tuberculosis: concluding analyses from a large-scale field trial', *International Journal of Infectious Disease* 11(4) (2007): 300–308, doi: 10.1016/j.ijid.2007.04.001). It found a roughly 25 per cent reduction in tuberculosis as a result of culling, but the study was flawed in a number of important ways. For example, it was impossible to blind the trial participants from which treatment (e.g. culling or no culling) was being used in different regions and some treatments were ended early when it appeared the results were going in the wrong direction. The study, although expensive and difficult to carry out, was also a weak test of the hypothesis that badgers were a cause of TB in cattle. The greatest flaw was, however, that the weaknesses of the study were initially overlooked when establishing how an operational TB control system could be designed. My role was mainly to shift the government to adopting an adaptive approach to the design of operational implementation in order to compensate for these weaknesses.

7 Eventually it took a ruling from the Supreme Court to jolt them into their senses about air quality.

Chapter 4 Rationalizing the politics factory

1 D. Graeber, *Debt: The First 5,000 Years*, Brooklyn, NY: Melville House, 2011, p. 387.

2 Lord Kelvin (Sir W. Thomson), 'Electrical units of meas-

urement', delivered 3 May 1883, in *Popular Lectures and Addresses*, vol. 1, *Constitution of Matter*, London: Macmillan, 1889, pp. 73–74 (original emphasis).

3 *Normal* politics is reflected here as a Machiavellian-Schumpeterian view of the world where, broadly, politics is a competition between those within the political classes rather than a representation of the balance of public opinion or a process for achieving a broad consensus.

4 For a discussion of scientific realism, see B.C. van Fraassen, *The Scientific Image*, Oxford: Clarendon Press, 1980, chapter 2.

5 W. Steffen et al., 'Planetary boundaries: guiding human development on a changing planet', *Science* 347 (2015): 1259855, doi: 10.1126/science.1259855.

6 *The Republic*, Book 7.

7 W. Blake, 'Proverbs of hell', in *The Marriage of Heaven and Hell* (1790).

8 J.K. Feibleman, *Understanding Philosophy: A Popular History of Ideas*, London: Souvenir Press, 1975.

9 The assumption that there are only two specific 'parties' representing democracy in the United States is illustrated by S.M. Sidik, 'How to tackle political polarization – the researchers trying to bridge divides', *Nature* 615(7950) (2023): 26–28, doi: 10.1038/d41586-023-00573-5.

Chapter 5 Gateways to the politics factory

1 S. Zuckerman, 'Science advisers and scientific advisers', *Proceedings of the American Philosophical Society* 124(4) (1980): 254.

2 G. Mulgan, 'COVID's lesson for governments? Don't cherry-pick advice, synthesize it', *Nature* 602 (2022): 9, doi: 10.1038/d41586-022-00212-5.

3 UK Government, *The Seven Principles of Public Life*, https://www.gov.uk/government/publications/the-7-principles-of-public-life.

4 R.M. Pirsig, *Zen and the Art of Motorcycle Maintenance: An*

Inquiry into Values, New York: Harper Torch, 2006 [1974], p. 139.

5 COBR is the Cabinet Office Briefing Room, a place where national emergency committee meetings are held in the UK.

6 See T. Dalyell, 'Politics and conservation: a view from inside Parliament', in I. Boyd and R. Crofts (eds), *Conserving Nature: Scotland and the Wider World*, Edinburgh: John Donald Publishers, 2005, pp. 165–173. The chapter was shadow-written by me based on conversations with Tam Dalyell.

7 Truth is defined here as a state of knowledge verifiable by multiple independent methods applied by disinterested observers.

8 This is also practised by governments, but increasingly this is being prevented by the gradual implementation of national statistical authorities, such as the Office for National Statistics in the UK.

9 The UK's COVID-19 Inquiry revealed the dissatisfaction there was among government Chief Scientific Advisers about the policies being adopted, expressed within diaries and WhatsApp messages.

10 P. Gluckman et al., 'Brokerage at the science–policy interface: from conceptual framework to practical guidance', *Humanities and Social Science Communication* 8 (2021): 84, doi.: 10.1057/s41599-021-00756-3. These authors also endorse and expand on Pielke's categorization of roles for science advisers (see R.A. Pielke, *The Honest Broker: Making Sense of Science in Policy and Politics*, Cambridge: Cambridge University Press, 2014). I am doubtful about this kind of reflection on the ways in which science advisers work. This may address the needs of social theory, but it does not reflect the instinctive ways in which science advisers actually have to operate.

11 Known as the *Kingdon Effect*. See J.W. Kingdon, *Agendas, Alternatives, and Public Policies*, 2nd edn, New York: HarperCollins, 1995.

12 UK Parliament, House of Lords Science and Technology Committee, 'Engagement with the policy process and departmental evidence quality assurance', in *The Role and Functions of Departmental Chief Scientific Advisers – Science and Technology Committee*, 2011, https://publications.parlia ment.uk/pa/ld201012/ldselect/ldsctech/264/26406.htm.

13 K. De Pryck and M. Hulme (eds), *A Critical Assessment of the Intergovernmental Panel on Climate Change*, Cambridge: Cambridge University Press, 2022.

14 Government Office for Science, 'Principles of scientific advice to government', 24 March 2010, https://www.gov.uk/govern ment/publications/scientific-advice-to-government-principl es/principles-of-scientific-advice-to-government.

15 For a short account of the role of the White House Office of Science and Technology Policy, see D.M. Hart, 'An agent, not a mole: assessing the White House Office of Science and Technology Policy', *Science and Public Policy* 41(4) (2014): 411–418, doi: 10:1093/scipol/sct061.

16 P. Gluckman, 'Policy: the art of science advice to government', *Nature* 507 (2014): 163–165, doi: 10.1038/507163a.

17 Various 'collaborations' exist to synthesize across many lines of enquiry. These include the Cochrane Collaboration, which focusses on healthcare, the Campbell Collaboration, which focusses on social policy, and the Juno Evidence Alliance, which focusses on agriculture.

18 Government Office for Science, *From Waste to Resource Productivity*, 14 December 2017, https://assets.publishing.ser vice.gov.uk/media/5a82dd9740f0b62305b94c13/from-waste -to-resource-productivity-final-report.pdf.

19 C.P. Snow, *Science and Government*, Cambridge, MA: Harvard University Press, 2013 [1960], p. 49.

20 C.S. Peirce, *The Essential Peirce: Selected Philosophical Writings*, 2 vols, ed. N. House et al., Bloomington: Indiana University Press, 1992–1998.

21 W. James, *Pragmatism: A New Name for Some Old Ways of*

Thinking, Cambridge, MA: Harvard University Press, 1975 [1907].

22 G.P. Zachary, *Endless Frontier: Vannevar Bush, Engineer of the American Century*, New York: Free Press, 1997, chapter 4.

23 Frederick A. Lindemann (Lord Cherwell) was Churchill's trusted science adviser through the Second World War, but the formal role of Chief Scientific Adviser in the UK was not constituted until 1964. Lindemann's role was controversial because many in the scientific community thought he tended to tell Churchill what he wanted to hear. Bob May considered Lindemann to be 'an indifferent scientist but skilled social climber' (foreword to Snow, *Science in Government*, p. viii). Lindemann was opposed in his proposal for carpet bombing of German cities by two of his contemporaries as scientific advisers, Henry Tizzard and Patrick Blackett. This was an issue which rankled with the scientific community long after the end of the Second World War.

24 Snow, *Science and Government*, p. 49.

Chapter 6 Shoring up the marriage

1 See a summary of the impact of the Rothschild Report in M. Parker, 'The Rothschild Report (1971) and the purpose of government-funded R&D – a personal account', *Palgrave Communications* 2 (2016):16053, doi: 10.1057/palcomms.2016.53.

2 OECD, *Science, Technology and Industry Policy Papers No. 21*, Paris: OECD Publishing, 2015, http://dx.doi.org/10.1787/5js33l1jcpwb-en.

3 The Joint Research Centre provides direct research support to the European Commission.

4 See D.G. Burnett, *The Sounding of the Whale: Science and Cetaceans in the Twentieth Century*, Chicago: University of Chicago Press, 2012.

5 A.M. Barandt, 'Inventing conflicts of interest: a history of

tobacco industry tactics', *American Journal of Public Health* 102(1) (2012): 63–71, doi: 10.2105/AJPH.2011.300292.

6   In 2023, it was revealed that 'since the late 1970s and early 1980s, ExxonMobil [had] predicted global warming correctly and skillfully' (G. Spuran et al., 'Assessing ExxonMobil's global warming projections', *Science* 379 (2023): 6628, doi: 10.1126/science.abk0063). While there is no evidence that ExxonMobil was involved in hacking emails, the company apparently chose not to publicize what it knew because its knowledge would have supported the views of climate scientists.

7   J. Rawls, *A Theory of Justice*, Oxford: Oxford University Press, 1971.

8   P. Singer, *Practical Ethics*, Cambridge: Cambridge University Press, 2006.

9   K. Huang et al., 'Veil-of-ignorance reasoning favors the greater good', *Philosophical Transactions of the National Academy of Science* 116(48) (2019): 23989–23995, doi: 10.1073/pnas.191012511.

10 P.D. Thacker, 'Conflicts of interest among the UK government's Covid-19 advisers', *British Medical Journal* 371 (2020): m4716, doi: https://doi.org/10.1136/bmj.m4716.

11 See K. Kupferschmidt, 'UK government should not keep science advice secret, former chief adviser says', *Science*, 11 May 2020, doi: 10.1126/science.abc7438.

12 R. Pielke, Jr and N. Lane, 'Memo for President Biden: five steps to getting more from science', *Nature* 587(7835) (2020): 546–547, doi: 10.1038/d41586-020-03148-w.

13 D. Bachtold, 'Arsenic victims to take British science body to court', *Science* 300(5622) (2003): 1066, doi: 10.1126/science.300.5622.1066a. Reported in A. Abbott and N. Nonsego, 'Italian seismologists cleared of manslaughter', *Nature* 515 (2014): 171, doi: 10.1038/515171a.

14 R. Lofstedt, 'The precationary principle in the EU: why a formal

review is long overdue', *Risk Management: An International Journal* 16(3) (2014): 137–163, doi.: 10.1057/rm.2014.7.

15 The World Trade Organization found that the European Commission had violated a trade agreement by, in effect, banning the import of GM foods for six years from 1998.

16 Global Commission on Evidence to Address Societal Challenges, *The Evidence Commission Report: A Wake-Up Call and Path Forward for Decisionmakers, Evidence Intermediaries, and Impact-Oriented Evidence Producers*, Hamilton: McMaster Health Forum, 2022, p. 103, https:// www.mcmasterforum.org/docs/default-source/evidence -commission/evidence-commission-report.pdf?sfvrsn=2fb92 517_11.

Chapter 7 The subjective by-pass

1 J.K. Feibleman, *Understanding Philosophy: A Popular History of Ideas*, New York: Horizon Press, 1986 [1975], p. 126 (original emphasis).

2 Y.N. Harari, *Sapiens: A Brief History of Humankind*, New York: Penguin Random House; Toronto: McClelland & Stewart, 2014, pp. 279–280.

3 The crisis of confidence in the standards applied in government is described by A.E. Lerman, *Good Enough for Government Work: The Public Reputation Crisis in America (And What We Can Do to Fix It)*, Chicago: University of Chicago Press, 2019. However, see J. Avorn, 'FDA standards – good enough for government work?', *New England Journal of Medicine* 353 (2005): 969–972, for a commentary on how the opposite applies to the standards used for approving new drugs.

4 F. Galton, 'Eugenics: its definition, scope, and aims', *The American Journal of Sociology* X(1) (1904): 1, doi: 10.1086/211280.

5 M. Nobles et al., 'Science must overcome its racist legacy: *Nature*'s guest editors speak', *Nature* 606 (2022): 225–227, doi: 10.1038/d41586-022-01527-z. Also see D.E. Willoughby,

Masters of Health: Racial Science and Slavery in US Medical Schools, Chapel Hill: University of North Carolina Press, 2022.

6 Note that I am only using Harari's words as an illustration of something which is very prevalent. I am not suggesting that Harari is the originator of this view.

7 M. Weber, *Economy and Society*, ed. and trans. K. Tribe, Cambridge, MA: Harvard University Press, 2019 [1921].

8 In 1918, Viscount Richard Haldane wrote a report on 'the machinery of government'. Although this was specific to the UK, it was probably one of the first expressions of how government needed to build science into its systems of decision-making. Haldane himself never expressed this as a *principle*, which became a convention, most probably promoted by the scientific community as a way of sustaining its own interests. The so-called *principle* has evolved from the very reasonable idea that scientists and the scientific community itself are in the best position to judge what constitutes quality in science, even when it is publicly funded. However, it has ended up being the idea that scientists alone should decide how to spend public money on science and, in an extreme form, that they should have relatively little accountability for how public good is generated from that funding. This perception has done science no favours at all when viewed from within the politics factory.

Chapter 8 Products of the politics factory – evidence: *quod erat demonstrandum*

1 P. Ayer, 'A new epistemology of evidence-based policy', *Policy & Politics* 48(2) (2020): 241–258, doi: 10.1332/030557319X15 657389008311.

2 N.L. Henry, 'Knowledge management: a new concern for public administration', *Public Administration Review* 34(3) (1974): 189–196, doi: 10.2307/974902.

3 *Data* are referred to in the plural. Whereas data can commonly be used in the singular, I have chosen here to use the plural form, mainly because it makes little sense to talk in

terms of 'a data' when referring to the plural terms of 'datum'. But I am also suspicious that some of the problems I see with how people use and perceive data arise because of the cognitive confusion between the use of the word *data* as a plural or singular noun.

4 C.A. Donnelly et al., 'Four principles to make evidence synthesis more useful for policy', *Nature* 558(7710) (2018): 361–364, doi: 10.1038/d41586-018-05414-4.

5 One of these processes, known as 'DELPHI', was used by the Rand Corporation to estimate the effectiveness of nuclear war tactics in the early 1960s (see N. Dalkey and O. Helmer, *An Experimental Application of the DELPHI Method to the Use of Experts*, The Rand Corportation, Santa Monica, Memorandum RM-727/1-Abridged, 1962, https://www.rand.org/content/dam/rand/pubs/research_memoranda/2009/RM727.1.pdf).

6 E. Gould et al., 'Same data, different analysts: variation in effect sizes due to analytical decisions in ecology and evolutionary biology', EcoEvoRxiv (2023), doi:10.32942/X2GG62.

7 A study of language used in the journal *Science* showed the frequency of hedging words such as 'might' and 'probably' declined by about 40 per cent from 1997 to 2021 (J. Brainard, 'In some scientific papers, words expressing uncertainty have decreased', *Science*, 28 July 2023, doi: 10.1126/science.adk0158).

8 I found in particular that the toxic subject of the environmental effects of pesticides created the circumstances where environmental pressure groups had influenced the research done to the extent that there was significant bias within the evidence (see A.M. Milner and I.L. Boyd, 'Toward pesticidovigilance', *Science* 357(6357) (2017): 1232–1234, doi: 10.1126/science.aan2683; and I.L. Boyd, 'An inside view on pesticide policy', *Nature Ecology & Evolution* 2(6) (2018): 920–921, doi: 10.1038/s41559-018-0557-8).

9 In November 2023, the charity Sense About Science began a

campaign to create a 'transparency standard in government' (https://www.justgiving.com/campaign/evidencestandard). In the UK, the Treasury *Green Book* attempts to codify these standards (*The Green Book: Central Government Guidance on Appraisal and Evaluation, 2022,* https://assets.publish ing.service.gov.uk/media/623d99f5e90e075f14254676/Green _Book_2022.pdf). The fact that Sense about Science felt that a campaign was needed reflects dissatisfaction with government's capacity to comply with its own standards.

10 Many advanced governments produce official statistics. These are very particular forms of evidence in the form of statistical facts which have been derived using rigorous methods.

11 An important additional note here is that it is essential not to assume that the responses given by people are an honest reflection of their preferences. Any survey would need to test for honesty in the respondents.

12 H. Nowotny et al., *Re-thinking Science: Knowledge and the Public in an Age of Uncertainty,* Cambridge: Polity, 2010. Also see C. Craig and S. Dillon, 'Storylistening's role in policy advice – response', *Science* 379(6638) (2023): 1198, doi: 10.1126/Science.adh4233.

Chapter 9 'What works' in the politics factory?

1 B. Russell, *History of Western Philosophy,* Hove: Psychology Press, 2004 [1946], p. 728.

2 The Obama White House, 'Find what works', https://obamaw hitehouse.archives.gov/administration/eop/sicp/initiatives/fi nd-what-works.

3 *https://www.gov.uk/guidance/what-works-network.*

4 The main report on this is: J.F. Bourne et al. 2007. *Final Report of the Independent Scientific Group on Cattle TB Presented to the Secretary of State for Environment, Food and Rural Affairs the Rt Hon David Miliband MP,* June 2007, https://www.bo vinetb.info/docs/final_report.pdf. See also C.A. Donnelly et al., 'Positive and negative effects of widespread badger culling

on tuberculosis in cattle', *Nature* 439 (2006): 843–846, doi: 10.1038/nature04454; and C.A. Donnelly et al., 'Impacts of widespread badger culling on cattle tuberculosis: concluding analyses from a large-scale field trial', *International Journal of Infectious Diseases* 11(4) (2007): 300–308, doi: 10.1016/j. ijid.2007.04.001.

5 For a summary of the contrasting view of Charles Sanders Peirce and William James, see J. Campbell, *William James, Charles Peirce, and American Pragmatism*, Ashland, OR: Knowledge Products, 2006 (CD audiobook).

6 See https://www.gov.uk/guidance/what-works-network.

7 J. Avorn, 'FDA standards – good enough for government work?', *The New England Journal of Medicine* 353(10) (2005): 969–972, doi: 10.1056/NEJMp058174.

8 A.M. Milner and I.L. Boyd, 'Toward pesticidovigilance', *Science* 357(6357) (2017): 1232–1234, doi: 10.1126/science. aan2683.

9 I.L. Boyd, 'An inside view of pesticide policy', *Nature Ecology & Evolution* 2 (2018): 920–921, doi: 10.1038/s41559-018-0557-8.

10 From *An Honest Liar* (2014), directed by Tylor Measom and Justin Weinstein. See https://www.imdb.com/title/tt224656 5/?ref_=tt_ch.

11 See J. Surowiecki, *The Wisdom of Crowds*, London: Abacus, 2005.

Chapter 10 Following the crowd

1 R.T. Lackey, 'Normative science', *Fisheries* 29(7) (2004): 38.

2 In my capacity as a Fellow of the Royal Society, I have often been asked to contribute to joint statements from global national academies on subjects such as food security or the marine environment. It is remarkable how aligned these national academy views are with the politics of their own national interests. Nothing from this experience has suggested to me that the scientists involved in composing these

statements are not deeply embedded in the politics factory within their home nations.

3 J. Ravetz, *Scientific Knowledge and Its Social Problems*, London: Transaction Publishers, 1996.

4 M. Strevens, *The Knowledge Machine: How Irrationality Created Modern Science*, New York: W.W. Norton, 2020.

5 F. Bacon, *Novum Organum* (1620), Aphorism XLI.

6 H. Littlecott et al., 'Effectiveness of testing, contact tracing and isolation interventions among the general population on reducing transmission of SARS-CoV-2: a systematic review', *Philosophical Transactions of the Royal Society* 381(2257) (2023): 20230131, doi: 10.1098/rsta.2023.0131

7 T.S. Kuhn, *The Structure of Scientific Revolutions*, 3rd edn, Chicago: University of Chicago Press, 1996.

8 K.R. Popper, *Conjectures and Refutations: The Growth of Scientific Knowledge*, London: Routledge, 1963.

9 C. Andrade, 'HARKing, cherry-picking, P-hacking, fishing expeditions, and data dredging and mining as questionable research practices', *The Journal of Clinical Psychiatry* 82(1) (2021): 20f13804, doi: 10.4088/JCP.20f13804.

10 S.J. Gould, *The Structure of Evolutionary Theory*, Cambridge, MA: Harvard University Press, 2002.

11 As an example, see a story carried by the BBC about how the methods used by manta rays to filter plankton might also be applied to filtering microplastics from water: 'Manta rays inspire new device to filter microplastics', BBC News, 27 September 2021, https://www.bbc.co.uk/news/av/science -environment-58573451.

12 B.G. Norton and R.E. Ulanowicz, 'Scale and biodiversity policy: a hierachical approach', *Ambio* 21(3) (1992): 244–249, doi: 10.1016/0006-3207(93)90515-3.

13 A.M. Barandt, 'Inventing conflicts of interest: a history of tobacco interest tactics', *American Journal of Public Health* 102(1) (2012): 63–71, doi: 10.2105/AJPH.2011.300292.

14 I found it was usually taken for granted by most researchers,

and even some research funding agencies, that this was true when, at the time, it was still an open, unanswered question.

15 A.W. Montford, *The Hockey Stick Illusion: Climategate and the Corruption of Science*, London: Stacey International, 2010.

16 See this argument about plastics in M.C. Rillig and A. Lehmann, 'Microplastic in terrestrial ecosystems', *Science* 368(6498) (2020): 1430–1431, doi: 10.1126/science.abb5979.

17 There is a difference between intentional research misconduct and systemic incentivization to construct research in ways which reify socio-political trends. An example involves the abnormal behaviour in the larvae of reef fish when exposed to acidified seawater. See M. Enserink, 'Sea of doubts', *Science* 372(6542) (2021): 560–565, doi: 10.1126/science.372.6542.560.

18 H.E. Longino, 'Theoretical pluralism and the scientific study of behavior', *Minnesota Studies in the Philosophy of Science* 19 (2006): 102–131, doi: 10.5749/j.cttttgnm.9.

19 C.E. Shannon, 'A mathematical theory of communication', *Bell System Technical Journal* 27(3) (1948): 379–423, doi: 10.1002/j.1538-7305.1948.tb01338.x.

20 See I.J. Good, 'On the principle of total evidence', *The British Journal for the Philosophy of Science* 17(4) (1967): 319–321, doi: 10.1093/bjps/17.4.319. Good concluded that it pays to take into account further evidence, provided that the cost of collecting and using this evidence, although positive, can be ignored. When this cost influences the reliability of the evidence itself, then this is likely to negate such a principle.

21 J.B. Carlisle, 'False individual patient data and zombie randomised controlled trials submitted to Anaesthesia', *Anaesthesia* 76(4) (2020): 472–479, https://doi.org/10.1111/anae.15263..

22 S.B. Nissen et al., 'Research: publication bias and the canonization of false facts', *eLife* 5 (2016): e21451, doi: 10.7554/eLife.21451.

23 I.E. Dror et al., 'Letter to the editor: The bias snowball and the bias cascade effects: two distinct biases that may impact forensic decision making', *Journal of Forensic Sciences* 62(3) (2017): 832–833, doi: 10.1111/1556-4029.13496.

24 Many of the crime scene dramas presented on television feed the public with a template for normative research.

25 M. Park et al., 'Papers and patents are becoming less disruptive over time', *Nature* 613 (2023): 138–144, doi: 10.1038/s41586-022-05543-x.

26 D. Graeber, *Bullshit Jobs: A Theory*, London: Allen Lane, 2018, p. 5.

27 Both the Cochrane and Campbell collaborations which build supervised methodologies for synthesizing research outputs include assessments for risk of bias and discount those studies where the risks are high.

28 H. Arendt, *Eichmann in Jerusalem: A Report on the Banality of Evil*, New York: Viking, 1963.

Chapter 11 Trust in experts?

1 'Gove: Britons "have had enough of experts"', interview with Faisal Islam of Sky News, 3 June 2016, available from https://www.youtube.com/watch?v=GGgiGtJk7MA.

2 To counter this, the European Union has begun moves to bring better discipline to how these analyses are conducted. See EU Competence Centre on Modelling and Decision Analysis, https://knowledge4policy.ec.europa.eu/modelling_en.

3 Public and private benefits have particular definitions. A private benefit exists when individuals can use something for their own, individual benefit, normally in the form of a marketable product which can be sold at a profit. Public benefit is when somebody provides a service or a product and expects no personal reward other than social recognition.

4 B. Obama, *A Promised Land*, New York: Viking, 2020, pp. 389–390.

5 E.O. Wilson, 'Scientists, scholars, knaves and fools', *American Scientist* 86 (1998): 7.

6 The 20th-century sociologists Robert Merton and Harriet Zuckerman named this the *Matthew Effect*, based on a parable from the Gospel of Matthew (R.K. Merton, 'The Matthew Effect in science', *Science* 159(3810) (1968): 56–63).

7 J. Brainard, 'Reviewers award higher marks when a paper's author is famous', *Nature* 377(6612) (2022): 1251, doi: 10.1126/science.ade8714.

8 I.L. Boyd, 'Coronavirus: a former member of SAGE explains how the science advisory group really works', *The Conversation*, 24 March 2020, https://theconversation.com /coronavirus-a-former-member-of-sage-explains-how-the -science-advisory-group-really-works-134077.

9 This specific example comes from Onora O'Neil's 2002 Reith Lectures, *A Question of Trust* (*https://www.bbc.co.uk/radio4 /reith2002/lecture1.shtml*).

10 W.J. Sutherland and M.A. Burgman, 'Policy advice: use experts wisely', *Nature* 526 (2015): 317–318, doi: 10.1038/526317a.

11 Foreword to C.P. Snow, *Science and Government*, Cambridge, MA: Harvard University Press, 2013 [1960], p. ix.

12 Climate models are also used to 'backcast', so they can suggest what the weather was fifty years ago. This projection can then be compared with the actual weather fifty years ago, resulting in an estimate of the accuracy of future projections.

13 Campaign for Science and Engineering, 'Public attitudes to R&D', https://www.sciencecampaign.org.uk/what-we-do/pub lic-opinion/public-attitudes-to-r-d/.

14 IPSOS Mori, 'It's a fact! Scientists are the most trusted people in world', 18 September 2019, https://www.ipsos.com/ipsos -mori/en-uk/its-fact-scientists-are-most-trusted-people -world.

15 S. Hoogeveen et al., 'The Einstein effect provides global evidence for scientific source credibility effects and the influence

of religiosity', *Nature Human Behaviour* 6 (2022): 523–535, doi: 10.1038/s41562-021-01273-8.

16 W.E. Bijker et al., *The Paradox of Scientific Authority: The Role of Scientific Advice in Democracies*, Cambridge, MA: MIT Press, 2009.

17 This includes DELPHI methods, which aggregate individual expert views independently of one another before then applying a deliberative method to build a consensus about the modal opinion.

Chapter 12 Redefining quality

1 S. Jasanoff, *The Ethics of Invention: Technology and the Human Future*, New York: W.W. Norton, 2016, p. 250

2 H. Else, '"Tortured phrases" give away fabricated research papers', *Nature* 596 (2021): 328–329, doi: 10.1038/d41586-021-02134-0.

3 H. Else, 'Paper-mill detector put to the test in push to stamp out fake science', *Nature* 612 (2022): 386–387, doi: 10.1038/d41586-022-04245-8.

4 R. Nuzzo, 'How scientists fool themselves – and how they can stop', *Nature* 526 (2015): 182–185, doi: 10.1038/526182a.

5 R.M. Pirsig, *Zen and the Art of Motorcycle Maintenance: An Inquiry into Values*. New York, Harper Torch, 2006 [1974].

6 O. Leyser, 'The excellence question', *Science* 370(6519) (2020): 886, doi: 10.1126/science.abf7125.

7 To sample this arrogance, I suggest reading books written by Peter Medawar.

8 'Editorial: Science needs to redefine excellence', *Nature* 554 (2018): 403–404, doi: https://doi.org/10.1038/d41586-018-02183-y.

9 J. Osborne and D. Pimental, 'Science, misinformation, and the role of education', *Science* 378(6617) (2022): 246–248, doi: 10.1126/science.abq8093. These authors propose a 'fast and frugal heuristic' for making a judgement about the quality

of scientific information which focusses on the credibility and expertise of the source.

10 L.I. Meho, 'The rise and rise of citation analysis', *Physics World* 20(1) (2007): 32–36, doi: 10.1088/2058-7058/20/1/33.

11 This refers to the manipulation of reference lists to enhance the reputations of specific scientists when judged using algorithms which reflect how often a piece of science is used by other scientists.

12 R. Van Noorden, 'Signs of "citation hacking" flagged in scientific papers', *Nature* 584(7822) (2020): 508, doi: 10.1038/d41586-020-02378-2.

13 E.M. Bik et al., 'The prevalence of inappropriate image duplication in biomedical research publications', *mBio* 7 (2016), doi: 10.1128/mBio.00809-16.

14 National Audit Office, *Good Practice Guide: Fraud and Error*, March 2021, https://www.nao.org.uk/wp-content/uploads/2021/03/010381-001-Fraud-and-Error-Accessible.pdf.

15 Association of Certified Fraud Examiners, *Report to the Nations: 2020 Global Study on Occupational Fraud and Abuse*, https://acfepublic.s3-us-west-2.amazonaws.com/2020-Report-to-the-Nations.pdf.

16 Retraction Watch, which monitors scientific papers which are retracted for reasons of inaccuracy or misconduct, has a league table showing people who have had most retractions. This table is dominated by men. See https://retractionwatch.com/the-retraction-watch-leaderboard/.

17 H. Wood et al., *The Silent Threat: The Impact of Fraud on UK National Security*, RUSI Occasional Paper, January 2021, https://static.rusi.org/the_silent_threat_web_version.pdf.

18 This is not the same as asking for an evidence standard, referred to in chapter 8. Since evidence is contextualized knowledge, quality in evidence would follow from ensuring the highest possible standards were applied to the people and processes used to generate knowledge as well as how that knowledge was interpreted and used. Other professions use

these kinds of standards, and they are especially prevalent in engineering.

Chapter 13 Playing the paradigm game

1 T.S. Kuhn, *The Structure of Scientific Revolutions*, 3rd edn, Chicago: University of Chicago Press, 1996, pp. 23–24.
2 Thomas Henry Huxley's opening address of the Fisheries Exhibition of 1882: https://mathcs.clarku.edu/huxley/SM5/fi sh.html.
3 Introduction to R.K. Merton, *The Sociology of Science: Theoretical and Empirical Investigations*, ed. N. Storer, Chicago: University of Chicago Press, 1979, p. xviii.
4 N. Eldredge and S.J. Gould, 'Punctuated equilibria: an alternative to phyletic gradualism', in T.J.M. Schopf (ed.), *Models in Paleobiology*, San Francisco: Freeman Cooper, 1972, pp. 82–115. Reprinted in N. Eldredge, *Time Frames: The Evolution of Punctuated Equilibria*, Princeton: Princeton University Press, 1985, pp. 82–115.
5 See chapter 4, note 3.
6 I say 'stupid' advisedly because there are a lot of people who seem to expect the ocean to produce to their unlimited demands.
7 Fisheries management is only one subset of the management of all natural resources upon which we rely for our existence, but the lessons are the same for everything we exploit, from critical minerals like uranium to oil, gas, coal, land for agriculture and even the water we drink and the air we breathe.
8 The World Organization for Animal Health (OIE) and, for plants, the International Plant Protection Convention (IPPC)
9 There was a gold standard until 1971, when US President Richard Nixon ended the convertibility of the US dollar to gold. The loss of the gold standard resulted in the US dollar becoming the standard, known as the international reserve currency. This was a very real paradigm shift in tokens of exchange because from having been pegged to

something tangible (gold), it was now based purely on trust. Cryptocurrencies and cryptoassets, such as non-fungible tokens, have arguably been the next paradigm shift where central banks, as the brokers of trust, were taken out of the equation. With cryptocurrencies, trust is vested in the integrity of a block chain, but experience has shown that the relative security this was supposed to bring has not materialised.

10 People like Peter Larkin were suggesting it had important shortcomings (P.A. Larkin, 'An epitaph for the concept of the maximum sustained yield', *Transactions of the American Fisheries Society* 106(1) (1977): 1–11, doi: 10.1577/1548-8659(1977)106<1:AEFTCO>2.0.CO;2).

11 Nevertheless, many scientists have created fudges which allow MSY to remain but where it is used in a highly precautionary form.

12 P. Dasgupta, *The Economics of Biodiversity: The Dasgupta Review*, London: HM Treasury, February 2021, https://asse ts.publishing.service.gov.uk/media/602e92b2e90e07660f80 7b47/The_Economics_of_Biodiversity_The_Dasgupta_Revi ew_Full_Report.pdf.

13 Y.N. Harari, *Sapiens: A Brief History of Humankind*, New York: Penguin Random House; Toronto: McClelland & Stewart, 2014.

14 In particular, Imre Lakatos and Larry Laudan modified the idea of paradigms (P. Godfrey-Smith, *Theory and Reality: An Introduction to the Philosophy of Science*, Chicago: University of Chicago Press, 2003, pp. 102–103).

15 My co-author Philippe Cury was the person who suggested the strap-line. Initially, I was not keen on such a potentially flippant effort, but I was wrong (P.M. Cury et al., 'Global seabird response to forage fish depletion – one-third for the birds', *Science* 334(6063) (2011): 1703–1706, doi: 10.1126/ science.1212928).

16 S. Dillon and C. Craig, *Storylistening: Narrative Evidence and*

Public Reasoning, Abingdon, Oxon: Routledge, 2021. Also see the discussion from P. Calow and S. Dillon, 'Storylistening's role in policy advice – response', *Science* 379(6638) (2023): 1198, doi: 10.1126/Science.adh4233; and C. Craig and S. Dillon, '"Storylistening" in the science policy ecosystem', *Science* 379(6628) (2023): 134–136, doi: 10.1126/science. abo1355.

17 See previous note.

Chapter 14 Taming wickedness

1 J.C. Camillus, 'Strategy as a wicked problem', *Harvard Business Review*, May 2008, https://hbr.org/2008/05/strategy -as-a-wicked-problem.

2 This can be presented as a classification of problems in four types or quadrants along the basic and applied axes: fundamental (or basic); relevant (applied); neither fundamental nor relevant; or both fundamental and relevant. This latter quadrant is known as Pasteur's quadrant, named after the great French chemist and microbiologist Louis Pasteur.

3 H.W.J. Rittel and M.M. Webber, 'Dilemmas in a general theory of planning', *Policy Science* 4 (1973): 155–169, doi: 10.1007/BF01405730. Rittel and Webber suggested wicked problems had ten properties, which included: they have no definitive formulation; they have no stopping rule; they have no true/false solutions; solutions are hard to recognize; each case is bespoke; the number of solution is not exhaustible; each problem is unique; each problem is a symptom of something else; each problem involves multiple options about what is right; and the person responsible is not permitted to be wrong.

4 It is worth noting that names like those listed for fields of science are in some senses themselves a political construct. At the time of writing, there is intense political interest in 'engineering biology', but this is a term largely invented to satisfy a short-term need for fads within the politics factory.

Those working on scientific research may engage with these clichés because they tend to determine how funding is distributed, but they are usually derived from (and therefore secondary to), rather than deterministic of, trends in thinking about problem-solving.

5 For a readable survey of complex problems, see M. Mitchell, *Complexity: A Guided Tour*, Oxford: Oxford University, 2009.

6 Lorenz originally used the image of a seagull flapping its wings, but his address to the American Association for the Advancement of Science on 29 December 1979 titled 'Predictability: does the flap of a butterfly's wings in Brazil set off a tornado in Texas?' has caught the imagination (E.N. Lorenz, *The Essence of Chaos*, Seattle: University of Washinton Press, 1993, pp. 181–184).

7 There seems to be a lot of confusion about where this quote originated. It is often attributed to Harold Wilson, but doubt is cast on that here: https://oupacademic.tumblr.com/post /66321507640/misquotation-a-week-is-a-long-time-in.

8 J.P. O'Dwyer, 'Stability constrains how populations spread risk in a model of food exchange', *One Earth* 2(3) (2020): 269–283, doi: 10.1016/j.oneear.2020.02.016.

9 C.W. Churchman, 'Guest editorial: Wicked problems', *Management Science* 14(4) (1967): B141.

10 M. Gerstung et al. 'The evolutionary history of 2,658 cancers', *Nature* 578 (2020): 122–128, doi: 10.1038/s41586-019-1907-7.

11 Elon Musk refers to the *idiot index* as the multiplier of cost, over and above the basic material costs, of manufacturing something like a rocket motor or a car (see W. Isaacson, *Elon Musk*, London: Simon & Schuster, 2023, pp. 99–100). It is a measure of the amount of inefficiency built in by the manufacturing process. Some economists would talk about this as 'added value', but Musk's point is that standard economic thinking is warped towards the creation of burgeoning inefficiency in the way we use and deploy limited resources. It

is wicked because it is maladaptive. Musk's view can be generalized as the *broken window fallacy* in economics, which states that the spending of money on things which are wantonly destroyed or are wasteful does not result in a net good. Much supply-side economics is constructed around this fallacy because of the ways it supports the rising consumption and destruction of resources. Fast fashion is a clear example, but it pervades most of what we buy and consume. While resources may be translated into liquid assets in the form of money by manufacturing, these financial assets are notional rather than real. The fallacy is revealed most starkly as the astronomical *idiot index* associated with fine art or fine wine, but everyday items like food and clothing also have high *idiot indexes* too. The reason this is wicked, or idiotic, is that the inefficiency multipliers created by this purely notional added value are then used to purchase access to even more natural resources. Because the *idiot index* is embedded in everyday items or services, it creates a runaway process of exploitation of resources which is hard to stop. Many liberal economic ideas have wickedness at their core.

12 Camillus, 'Strategy as a wicked problem'.

Chapter 15 Adaptive policy testing: making policy into science

1 H.G. Wells, *Mind at the End of Its Tether*, London: William Heinemann Ltd, 1945, p. 19.

2 R. Feynman et al., *The Feynman Lectures on Physics*, New Millenium Edtion, vol. 1, *Mainly Mechanics, Radiation, and Heat*, New York: Basic Books, 2010, p. 1-1.

3 Chris Tyler, when Director of the UK Parliamentary Office of Science and Technology, said 'there is no such thing as the policy cycle' (C. Tyler, 'Top 20 things scientists need to know about policy-making', *The Guardian*, 2 December 2013, https://www.theguardian.com/science/2013/dec/02/scientists -policy-governments-science).

4 In the UK, the guidance is contained within 'books' published by Her Majesty's Treasury. The *Magenta Book* provides guidance on how policy evaluation should be carried out, while the *Green Book* provides guidance on project, programme and policy appraisal, including the production of *impact assessments*, which have to be produced for any new policies.

5 In ethical theory this is often known as *the Trolley Problem* (J.J. Thompson, 'The Trolley Problem', *The Yale Law Journal* 94 (1985): 1395–1415, doi: 10.2307/796133).

6 The choice of the word 'kill' here is deliberate because the whole point is that we are actively killing people. Many environmental pollutants continue to be produced by people who have been told that by producing them they are killing people. Individually, our share of this killing may be small, but it is one of the greatest 'tragedies of the commons' (G. Hardin, 'The tragedy of the commons', *Science* 162(3859) (1968): 1243–1248, doi: 10.1126/science.162.3859.1243) that we are ill prepared to take responsibility for our own actions in this regard. We are not passive observers but active participants.

7 Utilitarian ethics sits at the centre of most neo-classical economic analyses, which are often the basis upon which governments work to make judgements about the balance between the costs and the benefits of specific kinds of policy interventions. It is frequently seen as much the same as consequentialism in ethics, but it suffers from the need to be able to have a reasonable framing of what consequences might result from specific actions. This brings in science as a process for projecting consequences. Economists often use this science but can sometimes underplay the uncertainty, leading to deceptive confidence in potential consequences.

8 I. Kant, 'Introduction', in *The Critique of Pure Reason*, trans. N. Kemp-Smith, London: Macmillan, 1929 [1781], B1, p. 41.

Chapter 16 More than just widgets

1 C.P. Snow, *Science and Government*, Cambridge, MA: Harvard University Press, 2013 [1960], p. 59.
2 This was addressed by John Ziman: *Prometheus Bound: Science in a Dynamic Steady State*, Cambridge: Cambridge University Press,1994; and 'Non-instrumental roles of science', *Science and Engineering Ethics* 9 (2003): 17–27, doi: 10.1007/s11948-003-0016-y.
3 The terms *public* and *private goods* derive from economics. A private good is something which is sold for a private or personal gain. A public good is something for which there is no direct personal gain and which is done to benefit wider society. All scientific research will have components of both, but in some areas of research one is likely to be stronger than the other. However, some people who run private businesses, such as farms, sometimes attempt to argue that they provide a public good. These are businesses aimed at personal gain, however, so they do not qualify as a public good.
4 F. Poege, 'Science quality and the value of inventions', *Science Advances* 5(12) (2019): eaay7323, doi: 10.1126/sciadv.aay7323.
5 The term *dual use* is usually used to denote technologies which can both be used within a market and be potentially weaponised.
6 A. Warhol, *The Philosophy of Andy Warhol (From A to B and Back Again)*, New York: Harcourt Brace Jovanovich, 1975, p. 229.
7 W.S. Jevons, *The Coal Question: An Inquiry Concerning the Progress of the Nation, and the Probable Exhaustion of our Coal-Mines*, ed. A.W. Flux, 3rd edn, New York: Augustus M. Kelley, 1905.
8 Electric light bulbs are classic widgets. The International Energy Agency noted that 'despite continued improvements in the efficacy of lighting', with their efficiency increasing at least 100 times since they were invented, 'increasing use of

lighting services drove up consumption, particularly in large emerging economies (IEA website: *https://www.iea.org/repor ts/lighting*).

9 D. Graeber, *Debt: The First 5,000 Years*, Brooklyn, NY: Melville House, 2011.

10 David Cameron, former British Prime Minister, and at the time President of Alzheimer's UK, was reported by the BBC as saying in response to the news of a successful clinical trial of the drug donanemab: 'We're a country of 60 million people, with a million people with dementia, many of them in very expensive residential care settings and so there is a lot of savings to be had from effectively treating people' (F. Walsh and M. Roberts, 'Drug donanemab seen as turning point in dementia fight', BBC News, 17 July 2023, https://www.bbc.co .uk/news/health-66221116).

11 The accountancy system is known as the 'ESA 10 rule'. It defines scientific research as an activity to build knowledge capital and places it on national balance sheets as an asset. The OECD uses the Frascati Definition to describe what counts as scientific research for the purpose of national accounting (Eurostat, *Manual on Measuring Research and Development in ESA 2010*, 2014 edn, *https://ec.europa.eu/eurostat/docu ments/3859598/5937049/KS-GQ-14-004-EN.PDF/eed4dfe2 -9b89-4c30-8c49-f6152912c1a7*).

12 Website for the UK Government Office for Science: https:// www.gov.uk/government/collections/foresight-projects.

Chapter 17 Fixing the marriage

1 Referring here to the Machiavellian-Schumpeterian type.

2 M. Henderson, *The Geek Manifesto: Why Science Matters*, London: Bantam Press, 2012.

3 In October 2023, the UK Secretary of State for Science, Innovation and Technology wrote to the Chief Executive of UK Research and Innovation (UKRI) to, in effect, order her to shut down an advisory committee two of whose members had

had tweeted views on a subject which she disagreed with. The event precipitated a threat by academics to withdraw their support for UKRI, the funding body which is tasked with distributing public funding for research and for providing strategic leadership in scientific research.

4 UK Statistics Authority: https://uksa.statisticsauthority.gov.uk /what-we-do/.

5 Office for Budget Responsibility: https://obr.uk/about-the-obr/what-we-do/

6 The ways in which scientific advisers within the government executive structure were conflicted was illustrated during the COVID-19 pandemic (D. Sidhar, 'I feared scientific advisers were being used by the government – the Covid inquiry shows they were', *The Guardian*, 3 November 2023, https://www. theguardian.com/commentisfree/2023/nov/03/scientific-advisers-covid-inquiry-chris-whitty-scientists?CMP=Share_iOSA pp_Other).

7 See the REDemo ideas presented by Giovanni Molteni Tagliabue: *'Rationalized and Extended Democracy': Inserting Public Scientists into the Legislative/Executive Framework, Reinforcing Citizens' Participation*, doi: 10.5281/ zenodo.5956997.

8 J. Rawls, *A Theory of Justice*, Oxford: Oxford University Press, 1971.

9 N. Zelic, 'Public reason, civic trust and conclusions of science', *Empirical Evidence and Philosophy* 69 (2018): 99–117, doi: 10.4000/estetica.3783.

10 I.L. Boyd, 'Saving the poor and vulnerable', *Science* 370(6515) (2020): 383, doi: 10.1126/science.abf2694.

Index

circular 137
consumption 18, 214, 266 n.11
efficiency 213, 269 n.8
experts and special powers
154–6
growth and competitiveness
4, 11, 17–18, 98, 137,
153, 213–14, 222, 224–5,
236–7 n.5(b)
indicators xii
liberal and neo-classical 266–7
n.11, 268 n.7
of Nature 32, 58
objectives 181, 188, 192, 215
as operational science 212
paradigms 181–2
policy 201, 226
supply- and demand-side 15,
213, 266–7 n.11
ecosystem *see* systems
Einstein, Albert 10
Eisenhower, Dwight D. 47–50
emissions 14–15, 53, 67, 136
see also air quality; pollution
empiricism *see* philosophy
Enlightenment 28, 52, 223, 236
n.27
environment xi, 12, 15, 51, 56–8,
119, 125, 135, 243 n.9,
256 n.2
awareness of 48
condition of 46, 142, 163, 179,
181, 183, 235 n.22
legislation 104
policy 48, 84, 98, 136, 201
pressure groups 48, 81, 144,
147, 254 n.8
environmentalism 178
Epicureanism 231 n3
epidemiology *or* epidemic 10,
157, 64
equity *or* equality 212, 229–30

error (statistical) 120–6, 131, 169
ESA 10 rule 270 n.11
Estonia 96
ethics x, 214, 231 n.3
dilemmas 9, 207
in economics 268 n.5
frameworks 208
Kantian 208
in politics 6, 36, 39, 62, 67, 73,
92
rationale 116
in science 6, 9, 47–8, 51–3, 60,
101, 102, 145, 214, 233
n.11
standards 115
utilitarian 208, 268 n.7
virtue 207
eugenics 114–15
Euler's number 36
European Environment Agency
51
European Food Standards
Authority (EFSA) 100
European legislation 104
European Union (EU) *or*
Commission 68, 90, 97–8,
100, 104, 153–5, 200, 222,
234 n.20, 250 n.3, 259 n.2
evidence xiv, xvi, 38, 40
bogus 88
calls for 117, 42
construction of 93, 113
corruption of 117–65, 226
dismissal of 49
legal or forensic 74–5, 132, 150
policy based on xvi, 80, 117
scientific 104–5, 223, 255 n.10
selectivity or dismissal of 49,
81
as a social construct 118, 129
standards 127, 262 n.18
support system 105